C000254030

Defeat and Memory

Also by Jenny Macleod

GALLIPOLI: Making History (*editor*)

UNCOVERED FIELDS: Perspectives in First World War Studies
(*co-editor with Pierre Purseigle*)

Defeat and Memory

Cultural Histories of Military Defeat in the Modern Era

Edited by
Jenny Macleod
Lecturer in Twentieth-Century History, University of Hull

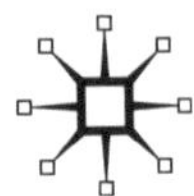

First published 2008 by
PALGRAVE MACMILLAN

Palgrave Macmillan in the UK is an imprint of Macmillan Publishers Limited, registered in England, company number 785998, of Houndmills, Basingstoke, Hampshire RG21 6XS.

Palgrave Macmillan in the US is a division of St Martin's Press LLC, 175 Fifth Avenue, New York, NY 10010.

Palgrave Macmillan is the global academic imprint of the above companies and has companies and representatives throughout the world.

Palgrave® and Macmillan® are registered trademarks in the United States, the United Kingdom, Europe and other countries.

ISBN-13: 978-0-230-51740-0 hardback
ISBN-10: 0-230-51740-4 hardback

This book is printed on paper suitable for recycling and made from fully managed and sustained forest sources. Logging, pulping and manufacturing processes are expected to conform to the environmental regulations of the country of origin.

A catalogue record for this book is available from the British Library.

Library of Congress Cataloging-in-Publication Data

Defeat and memory : cultural histories of military defeat in the modern era / [edited by] Jenny Macleod.
p. cm.
Includes bibliographical references and index.
ISBN-13: 978-0-230-51740-0 (hbk. : alk. paper)
ISBN-10: 0-230-51740-4 (hbk. : alk. paper) 1. Military history, Modern. 2. Collective memory. 3. Defeat (Psychology) I. Macleod, Jenny.

D214.D44 2008
355.0209'04—dc22 2008025122

10 9 8 7 6 5 4 3 2 1
17 16 15 14 13 12 11 10 09 08

Printed and bound in Great Britain by
CPI Antony Rowe, Chippenham and Eastbourne

Contents

Preface

This book developed from a conference that was hosted by the School of History and Classics, University of Edinburgh, in September 2005. The conference was facilitated by the sponsorship of the School of History and Classics, the British Academy, the Daiwa Anglo-Japanese Foundation, and the German History Society. I would like to take this opportunity to thank those who helped to organize the conference: Professor Jim McMillan, Dr Paul Addison, Dr Jeremy Crang, Dr Donald Bloxham, Professor Jill Stephenson, Seumas Spark, and the marvellous Mrs Pauline Maclean. Dr Andy Webster helped me to develop the original concept for the conference. Professor John Horne, Professor Dennis Showalter, and Dr Pierre Purseigle have provided me with invaluable advice, and friends and colleagues in Edinburgh and Hull have supported the process of editing this book. My final thanks go to all the speakers and participants at the conference, and in particular to the contributors to this book for their cooperation and commitment.

JRM

Notes on Contributors

Donald Bloxham is Professor of Modern History at the University of Edinburgh. His publications include *Genocide on Trial: War Crimes Trials and the Formation of Holocaust History and Memory* (Oxford: Oxford University Press, 2001) and *The Great Game of Genocide: Imperialism, Nationalism and the Destruction of the Ottoman Armenians* (Oxford: Oxford University Press, 2005).

Karen L. Cox is an Associate Professor and Director of Public History at the University of North Carolina at Charlotte. Her publications include *Dixie's Daughters: The United Daughters of the Confederacy and the Preservation of Confederate Culture* (Gainesville, FLA: University Press of Florida, 2003).

Kevin Cramer is an Associate Professor of History at Indiana University Purdue University Indianapolis. His publications include *The Thirty Years' War and German Memory in the Nineteenth Century* (Lincoln, NE: University of Nebraska Press, 2007).

Patrick Finney is a Lecturer in International Politics at the University of Wales, Aberystwyth. He is the editor of *Palgrave Advances in International History* (Basingstoke: Palgrave Macmillan, 2005).

Madoka Futamura is an Academic Programme Officer at Peace and Governance Programme, United Nations University. Her publications include *War Crimes Tribunals and Transitional Justice: The Tokyo Trial and the Nuremberg Legacy* (London: Routledge, 2008).

John Horne is Professor of Modern European History at Trinity College, Dublin. He is the co-author (with Alan Kramer) of *German Atrocities, 1914: A History of Denial* (New Haven: Yale University Press, 2001).

Jeffrey Kimball is Professor Emeritus at Miami University, Ohio. His publications include *The Vietnam War Files: Uncovering the Secret History of Nixon-Era Strategy* (Lawrence, KAN: University Press of Kansas, 2004).

Christian Koller is a Senior Lecturer in Modern History at Bangor University. His publications include *'Fremdherrschaft': Ein politischer Kampfbegriff im Zeitalter des Nationalismus* (Frankfurt: Campus, 2005).

Jenny Macleod is a Lecturer in Twentieth-Century History at the University of Hull. Her publications include *Reconsidering Gallipoli* (Manchester: Manchester University Press, 2004).

M. G. Sheftall is an Associate Professor of Communication Studies in the Faculty of Informatics at Shizuoka University. His publications include *Blossoms in the Wind: Human Legacies of the Kamikaze* (New York: NAL Caliber, 2005).

Anatol Shmelev is a Research Fellow at the Hoover Institution, Stanford University. He is the editor and compiler of the second edition of *The Gering Bibliography of Russian Émigré Military Publications* (New York: Ross Publications, 2007).

Stephen Tyre is a Lecturer in History, St Andrews University. His publications include '*Algérie française to France musulmane*: Jacques Soustelle and the Myths and Realities of "integration", 1955–62' in *French History*, Spring 2006.

Karine Varley is a Lecturer in History at the University of Edinburgh. Her publications include *Under the Shadow of Defeat: The War of 1870–1 in French Memory* (Basingstoke: Palgrave Macmillan, 2008).

Vanda Wilcox is an Adjunct Professor of Modern History at John Cabot University, Rome. Her publications include 'Discipline in the Italian Army 1915–1918' in Pierre Purseigle (ed.), *Warfare and Belligerence* (Leiden: Brill, 2005).

1
Introduction

Jenny Macleod

From Appomattox Courthouse to the Hall of Mirrors at the Versailles Palace, from the beaches of Dunkirk to the toppling of the statue of Saddam Hussein in Baghdad, the moment of victory in any conflict, whether temporary or definitive, has and will always incorporate an equally powerful moment of defeat. History may be written by the victors, but it is lived equally by the vanquished. The aim of this collection is to juxtapose a series of detailed case studies of the manner in which military defeat has been understood by individuals and societies in the era of modern industrialized warfare.

The defeats discussed here range from the Napoleonic Wars to the wars of late twentieth century. It is submitted here that, firstly, the memory of defeat provides a powerful prism through which to view modern history. Furthermore, that woven through these diverse examples are ideas and patterns of behaviour in response to defeat that sometimes echo each other and thus offer further insights in aggregate. This analytical framework has seldom been used previously. The exception is Wolfgang Schivelbusch's *The Culture of Defeat: On National Trauma, Mourning and Recovery*.[1] His book encompasses three defeats within six decades: the American South after the Civil War, France after its defeat at the hands of Prussia, and Germany after the First World War. In identifying the three sustaining ideas which shaped the memory of these defeats – the Lost Cause, revanchism, and *im Felde unbesiegt* (undefeated on the field of battle) – Schivelbusch locates these particular defeats within a wider pattern of responses. He notes the cyclical quality of victory and defeat, the challenge to complacency and hence the intellectual inspiration that defeats have provided, and the transference of blame to deposed tyrants and corrupt former regimes. Further recurring responses are the accusations of betrayal directed both internally and externally, the characterization of opponents as unworthy and unsoldierly and hence the celebration of the purity of the defeated, and the desire for revenge engendered by the humiliation.[2] Clearly, this is a theme that is rich in its potential and has a great deal of scope for the further investigation of the aftermath of other military humiliations. Although this collection was not conceived as a response to Schivelbusch, it develops and sometimes counters some of his ideas. In particular, Patrick Finney argues that the 'consolatory tropes' employed by the defeated in Schivelbusch's case studies were of limited utility in the wake of the Second World War. This strengthens the case for further investigation. Four chapters in this collection, including Finney's, take up this challenge. In addition, Jeffrey Kimball engages more directly with Schivelbusch who has cited Kimball on the matter of the myth of betrayal regarding the Vietnam War.

If the cultural history of defeat is relatively novel as a field of enquiry, the study of specific defeats is not. Many of them have approached the subject from the standpoint of political and military history as an analysis of what happened, what went wrong, and what the consequences were in terms of institutions and doctrines. As elsewhere within the discipline, more recently, historical analyses of different defeats have also incorporated elements of cultural history through historiographical assessments and reflections on the memory of the defeat,[3] or have solely taken this approach.[4] Sometimes military historians have chosen to analyse a series of defeats with a view to extracting more generalized lessons from the failings of the armed forces concerned.[5] Political scientists have taken this a step further in studying conflict termination and using case studies as a means to develop theories as to why and how wars end at one particular moment.[6] Thus, while the concept of defeat has become a subject of enquiry itself in strategic studies and cultural histories of individual defeats have been written, this collection is a contribution towards developing a cultural history of defeat itself.

In doing so, it rests not only upon earlier case studies but also on the flowering of the cultural history of war in the last two decades or so. Central to this are the ideas which have filtered down from the *Annales* school's interest in mentalities via Pierre Nora's grand investigation of sites of memory.[7] The exploration of the way in which war has been perceived and remembered – whether individually or, more problematically and amorphously, collectively – has enriched military history and broadened the sources through which historians approach their subject. Perhaps most commonly, acts of mourning and commemoration have been a particular focus of research. They are represented in this collection through the study of war memorials, ossuaries, battlefield pilgrimages, and commemorative ceremonies. But beyond this, there is a rich variety of sources where defeat has left a trace. The essays presented here draw upon historical interpretations ranging from unit histories, professional histories (consciously revisionist or otherwise) and school textbooks, museum collections, and the work of tribunals, as well as diverse elements of high and popular culture, such as paintings, popular songs, émigré newspapers, advertisements, and *manga* comics as a means to excavate attitudes to past defeats. This range of sources helps to mark out the distinctive contribution that cultural history can make. They enable the cultural historian to access ideas and issues that remain unaddressed by other disciplines.

The collection begins with a wide-ranging essay by John Horne which explores the importance of defeat in modern Western history and some of the patterns of responses which have followed. He argues that defeats that have befallen nation states have resounded more powerfully than earlier defeats might have done. This is because nation states have often been reliant on national mobilization for the war effort while being bounded by a strong sense of geographical integrity and replete with institutions that can process the repercussions of defeat. He delineates five different types of defeat which have been used to organize the chapters in this book. The first category in Horne's typology is temporary defeat which is later reversed in the overall outcome of the war. Examples featured here are the Prussian defeat by Napoleon in the double battle of Jena and Auerstedt in 1807 and the Italian defeat at Caporetto at the hands of a German-Austrian force in October 1917 on the Isonzo front in the First World War. The second category, definitive defeat, which Horne describes as 'the final verdict of a war that shapes the subsequent peace', is represented here in a chapter on the French memory of the Franco-Prussian War of 1870. The legacy of a total defeat, the third of Horne's categories, is explored in a number of chapters on the processing of the German and Japanese defeats in 1945 in the Second World War. The chapter on the way in which the modern memory of the Thirty Years' War in the seventeenth century informed German attitudes to sacrifice and annihilation in war in the twentieth century can also be located here. The fourth category, internal defeat, is reflected here in two chapters of contrasting scale. One of them is concerned with aspects of the overall legacy of the Confederate defeat in the American Civil War (1861–5), while the other examines the fate of the remnants of the Russian White Army defeated in 1920 by the Red Army in the Russian Civil War. The final category is that of partial defeat where there is no threat to the territory of the defeated nation. Often these are colonial defeats. The penultimate chapter examines the relationship between the French defeat at Dien Bien Phu in Indochina in 1954 and the subsequent Algerian War of Independence (1954–62). The French withdrawal from Indochina led to the deepening involvement there of the United States. This collection thus concludes with a chapter on the American memory of defeat in Vietnam which was sealed with the fall of Saigon to nationalist North Vietnam in 1975 some two years after the withdrawal of American combat forces.

In aggregate, these essays demonstrate that the humiliation of defeat was experienced simultaneously at the personal, institutional, and national levels. Two types of individuals are particularly visible here: the veteran and the scapegoat. Stephen Tyre's chapter explores the way in which Dien Bien Phu shaped its veterans' perceptions of their subsequent involvement in Algeria, framing it as part of a wider story of French decline and disaffection with the Fourth Republic. Thus personal and national aspects of the humiliation became inextricably entwined. M. G. Sheftall has studied the fate of the thousands of would-be *kamikaze* pilots who had been trained for suicide missions that would ensure their status as war gods only to find that Japan's surrender meant the profoundest of anticlimaxes. Sheftall contrasts two veterans' groups. One of them has been particularly concerned with personal ties and the honouring of their dead comrades; the other propounds a more politically and theologically charged interpretation of the *kamikaze* dead as emblematic of a heroic Japanese masculinity. Anatol Shmelev's chapter is also concerned with a veterans' association which sustained an idea of a 'morally and spiritually superior warrior'. In this case the association is dedicated to I Army Corps of the White Russian Army under the command of Lieutenant-General Kutepov which was interned on the Gallipoli peninsula, not far from the site which coincidentally has nurtured a more powerful and long-lasting myth of defeat, the Anzac legend. For their part, the defeated Russian army drew back from demoralization and defeatism to a dream of redemption and a renewed fight one day against the Bolsheviks. It was a dream that died with the individual veterans, as each one's passing was noted in the journals dedicated to the increasingly widely dispersed family of the defeated.

Thus armies, even disbanded ones of ageing veterans, can be the means by which memories are sustained. Indeed armies could be seen as the critical instrument whereby the individual, collective, and national levels of the experience of defeat are mediated. For example, the army as an institution can in some instances become the focus for a patriotic cult after it has suffered a defeat. As Karine Varley argues in her chapter, in the wake of the Franco-Prussian War, the republic celebrated the army's devotion to the nation as a means to look beyond the divisions that had scarred the country. Elsewhere, the search for individual scapegoats within the military could be superseded by the ideologically charged veneration of the military. Vanda Wilcox's chapter on the Italian defeat at Caporetto traces the way in which the commander-in-chief, Cadorna, blamed the rank and file before being blamed himself along with the Supreme Command by a Royal Commission which found itself unable to criticize members of the government that had set it up. After 1922, however, the army's circumstances changed as the fascist project required the participation of the military. Consequently, fascist rhetoric regarding Caporetto celebrated both commanders and commanded, and instead denounced the failings of the liberal government.

In other circumstances, the military has been more formally pitted against the civil power in the search to identify those responsible. Madoka Futamura shows here that in the International Military Tribunal for the Far East (the Tokyo Trial), the judgement focused on a military clique to the exclusion of key civilian figures and the Emperor in particular. The International Military Tribunal for the Trial of German Major War Criminals (the Nuremberg Trial) also differentiated between institutions. The Nazi Party, SS (*Schützstaffel*), *Gestapo*, and Hitler Youth were declared criminal organizations, but the *Wehrmacht* was not. Donald Bloxham's chapter encompasses not just the Nuremberg and Tokyo trials but also their equivalents in Turkey after the First World War and the trial of Slobodan Milosevic at The Hague. He shows that the US Chief Prosecutor Robert H. Jackson sought to differentiate between Nazis and ordinary Germans. During the trial, the German public expressed sympathy with the service chiefs whose responsibility for war crimes was of a different order than that of the highly politicized defendants, and outrage at the court's rejection of obedience to orders as a defence. Thus defunct institutions along with key individuals, took the blame, thereby in the immediate aftermath of the war shielding the rest of the country from the full burden of collective responsibility.

Let us now turn to look at five recurring themes that are threaded through the chapters gathered here: humiliation and honour, sacrifice and renewal, the stab-in-the-back, the Lost Cause, and taboo and collective amnesia.

Defeat is often marked by ritual. Sometimes the moment of surrender may be leavened by the dignity of the vanquished and the respect accorded to them. At the end of the American Civil War, General Robert E. Lee arrived in full-dress uniform with sash and jewelled sword to surrender and three days later when his troops surrendered their arms and their flags, the Union soldiers honoured them by shifting their stance to carry arms.[8] Sometimes defeats can even be celebrated as heroic events, as in the Australian and New Zealand memory of Gallipoli or the British celebration of the plucky spirit of the Dunkirk evacuation. More often, the rituals of defeat denote profound humiliation. Both Karine Varley and Christian Koller begin their chapters by observing the profoundly humbling consequences of defeat. For France in 1871, this meant lost territory, heavy indemnities, and the victors' triumphal march through the capital, for Prussia after 1807, significant territorial losses and foreign occupation. In these moments, the grievous physical loss of men and destruction of armaments on the battlefield commingle with economic debilitation, geographical mutilation, and political division and disempowerment in a profound psychic wound. The humiliation of defeat, be it personal or national, is an implicit thread running through all of these chapters, and the remaining themes identified here all arise from the need to assuage the pain it caused.

The idea of sacrifice leading to redemption lies at the heart of Christian theology and, arguably as Schivelbusch notes, has even deeper antecedents in Western culture in the myth of Troy.[9] It proved a powerful message for a number of predominantly Christian countries featured here. Kevin Cramer argues that nineteenth-century histories of the Thirty Years' War were framed within notions of regeneration and renewal, and that this remained a potent idea for German nationalists during and after the First World War. It helps to explain both the almost irrational last gasp offensives of 1918 and the conception of the search for Lebensraum as another religious war. In these cases, martyrdom and annihilation held within them the possibility of renewal. Koller's study of the memory of Jena and Auerstedt fits within the broader chronological framework of Cramer's piece, and presents a complementary argument. The defeat and the foreign rule which followed, he argues, were fitted into the classical structure of paradise, fall, purification, and rebirth. It was agreed that 1806 was the low point, but the main area of debate was whether the rebirth had yet occurred. After unification in 1871, the process was widely seen to have been fulfilled. Elsewhere, for example in Italy, as Wilcox shows, after an initial period of recrimination, Caporetto came to be interpreted as a trial of faith. The response to it was thus a moment of national renewal; hence the incorporation of images of redemptive sacrifice in the memorialization of the battle. Similarly, the White Russian veterans of internment at Gallipoli described their experiences in terms of suffering and transfiguration.

One of the most famous interpretive myths of defeat is the idea of the stab-in-the-back in which Germany's failure in the First World War was presented not as a result of military shortcomings, but because the home front had not properly supported the war effort. It finds a more recent echo in the Vietnam War, discussed here by Jeffrey Kimball. The idea reflects profound tensions between the military and civilians, and more specifically, in both Germany and America, appears to be a favoured weapon of the political Right. Kimball lists those blamed for Vietnam: 'leftists, liberals, the press, the antiwar movement, civilian policymakers, Democratic Party presidents, and the Congress of the United States'. He also finds that the idea chimes with that of the Lost Cause, the notion that became attached to the American South after the Civil War. In circumstances where the army had been disbanded, the Lost Cause can nonetheless be seen as a patriotic cult of the army in which its veterans are remembered as heroes and its leaders as martyrs. It served to reshape the cause of the war as pertaining to the honourable idea of states' rights, rather than that of slavery. As Karen Cox's chapter shows, paradoxically, the Lost Cause's celebration of the anti-modern features of the chivalrous South was perpetuated through the modern advertising techniques of mass consumerism. Nor is the Lost Cause solely an idea that appeals within America: M. G. Sheftall also identifies a desire within Japanese revisionism at one stage to establish the *kamikaze* as a Lost Cause in Japan.

More commonly, outsiders have perceived Japan as responding to its defeat in the Second World War through a convenient bout of historical amnesia. Patrick Finney locates the origins of this in the nature of the post-war settlement imposed by America in the context of the Cold War. The consequent dominance of the Liberal Democratic party resulted in a conservatism that favoured the more palatable explanations of Japan's role in the war (although it never entirely silenced some 'voices of conscience'). By contrast, Madoka Futamura characterizes Japan's memory of the war as refracted through the Tokyo Trial as more marked by taboo than amnesia – it was unmentionable rather than forgotten. She highlights three issues at the heart of this social prohibition. Firstly, the trial prosecuted Japan for waging an aggressive war and this contradicted the newly constructed image of Japan as a peace-loving nation. Secondly, it was feared that an unduly close examination of the trial's subject matter might reveal the Emperor's guilt. Most significantly, the trial offered an unsettlingly specific account of a war which is often blandly characterized as 'the previous war' in order to veil the absence of agreement on the nature of the war itself. Taboos and amnesia are a response to elements within the experience of defeat that are too difficult to deal with. They are by no means unique to Japan, of course. Finney argues that Italians repressed memories of the fascist dictatorship, and comforted themselves with a self-image that was decent if unheroic. Similarly, the myth of resistance in France took hold in a period when the true nature of Vichy was forgotten. Such myths were coming under question by the 1970s, at which point, Stephen Tyre argues, Algeria became the new Vichy insofar as it was France's best-forgotten war. Earlier, the unmentionable element of the memory of the Franco-Prussian War was the Paris Commune and its story of bloody civil war. As this and other examples show, although historians may instinctively view amnesia as reprehensible and seek to fill the lacuna, Karine Varley argues (citing Nietzsche and Renan) that amnesia in the face of profound trauma can be healthy both for an individual and for the sake of national unity.

Beyond the new light that is shed on a range of national histories by these cultural histories, they also offer some insights into the nature of history and memory. The first is the longevity of the influence that these defeats can have. The chapter with the longest time frame here concerned the influence of a seventeenth-century war on the mindset of the twentieth century in Germany – such was the scorching nature of the Thirty Years' War. Perhaps the most short-lived memory was that of the defeated White Russians, whose quest for honour through the Gallipoli Society slowly died away with the veterans to whom it had personally mattered. The most recent of the defeats featured here, Vietnam, is scarcely more than 30 years old at present. That its memory is still actively being processed is perhaps reflected in the politicized nature of this chapter.

The second point follows on from this: the mutability of the memory of these defeats. Patrick Finney's chapter – if we accept the work of professional historians as an act of collective remembrance[10] – vividly demonstrates the interaction between political context and interpretations of the defeats which befell the Axis powers in 1945. The onset of the Cold War made a vital difference, but so did the 1960s which brought generational shifts as well as changes within the discipline and practical advantages like the new availability of primary sources. In some cases, most obviously here in Cramer's view of the Thirty Years' War, the mutability of memory ensured its longevity. But eventually changed circumstances will limit the shelf life of a memory, especially after a total defeat such as Germany experienced in 1945 which, as Koller states, rendered dysfunctional earlier memories of defeat which promised renewal through further war.

What all of these comments point to is the importance of a usable past. Defeat in battle has the capacity to take on a significance beyond a mere judgement on military efficiency; it can seem to reflect on the defeated society itself. As such, finding a way to understand this humiliating event can prove profoundly challenging. These chapters show the ebb and flow of remembering and forgetting in the quest to come to terms with the past. Sometimes, as W. G. Sebald has shown, the task is impossible, the past so intolerable; it must be ignored and forgotten.[11] But this seems to be a temporary phenomenon when it occurs, and the work of processing a traumatic memory must begin at some point. These chapters suggest the variety of implications that this process can have. From the equanimity of individuals affected by the defeat to the tenor of international relations, the memory of defeat can leave a trace in a multitude of ways. The interpretation of defeat thus serves a vital social and political function in explaining and, perhaps, ameliorating the humiliation. The work of historians is central to this task, and it is hoped that this collection inspires further enquiry in this rich vein.

Notes

1. Wolfgang Schivelbusch, *The Culture of Defeat: On National Trauma, Mourning, and Recovery*, trans. Jefferson Chase (London: Granta, 2003).
2. Ibid., p. 1–35.
3. See, for example, the contrasting approaches of Alistair Horne, *To Lose a Battle: France 1940* (London: Macmillan, 1969) and Julian Jackson, *The Fall of France: The Nazi Invasion of 1940* (Oxford: Oxford University Press, 2003), especially pp. 185–249.
4. The seeds of the idea for this collection, for example, lie in my doctoral research on the memory of Gallipoli: Jenny Macleod, *Reconsidering Gallipoli* (Manchester: Manchester University Press, 2004). Other examples include Gaines M. Foster, *Ghosts of the Confederacy: Defeat, the Lost Cause, and the Emergence of the New South, 1865 to 1913* (New York: Oxford University Press, 1985); Carol Reardon, *Pickett's Charge in History and Memory* (Chapel Hill: University of North Carolina Press, 1997); and Jean-Marc Largeaud, *Napoléon et Waterloo: la défaite glorieuse de 1815 à nos jours* (Paris: La Boutique de l'Histoire editions, 2006).
5. See, for example, Norman F. Dixon, *On the Psychology of Military Incompetence* (London: Jonathan Cape, 1976); E. A. Cohen and J. Gooch, *Military Misfortunes: The Anatomy of Failure in War* (New York: Free Press, 1990); Brian Bond (ed.), *Fallen Stars: Eleven Studies of Twentieth Century Military Disasters* (London: Brassey's, 1991).
6. M. I. Handel, *War, Strategy and Intelligence* (London: Frank Cass, 1989).
7. P. Nora, *Realms of Memory*, 3 vols, ed. and trans. A. Goldhammer (New York: Columbia University Press, 1996–8).

8. James M. McPherson, *Battle Cry of Freedom: The Civil War Era* (New York: Oxford University Press, 2003), pp. 836, 850.
9. Schivelbusch, *The Culture of Defeat*, p. 2.
10. Jay Winter discusses the distinction between the two and the way in which history and memory are 'braided together' in *Remembering War: The Great War between Memory and History in the Twentieth Century* (New Haven: Yale University Press, 2006).
11. W. G. Sebald, *On the Natural History of Destruction*, trans. Anthea Bell (London: Hamish Hamilton, 2003), p. viii.

2
Defeat and Memory in Modern History

John Horne

Defeat looms large in memory, and thus history, because it marks rupture and renewal even more obviously than its inescapable twin, victory. Yet a literature search reveals little that addresses defeat and its memory in a thematic way. The obvious exception to this generalization is the fine study by Wolfgang Schivelbusch, *The Culture of Defeat*, which compares the American South after the Civil War, France after 1870 and Germany after 1918.[1] This apart, most studies of the subject are of individual defeats and are unreflective of the phenomenon as a whole. The task is enormous, even limited to the period since the Revolutionary and Napoleonic Wars. My ambition here is simply to establish some distinctions that may be useful in thinking further about how comparisons between different individual defeats might be framed. I shall look first at different types of defeat and then at what, if anything, distinguishes defeat in the last two centuries from earlier periods. Finally, I shall consider the relationship between defeat as experience and as memory.

Types of defeat

Although the subject of this book is the memory of defeat, its context is the nature of defeat – its shape, causes and consequences. For how defeat is remembered, and also forgotten, is partly determined by the kind of defeat it is. So we might begin by distinguishing some different types of defeat. Five come to mind.

The first is temporary defeat in a battle or campaign that is later reversed by the outcome of war. While this might seem less significant than definitive defeat, some of the most important defeats have been of this kind. The humiliation of Prussia in 1806 at Auerstadt and Jena illustrates the point. The period of internal reform in which the Prussian monarchy first grappled with the implications of the French Revolution, the victory at Leipzig in 1813 that followed and the national mythology that was invested in the 'War of Liberation' were all responses to this earlier defeat whose influence continued to shape Prussian and German history into the twentieth century.[2] The 1812 invasion of Russia might also be included in this category. There was no military defeat (the Battle of Borodino was a stand-off), but the abandonment of Moscow without a struggle and then its destruction by fire were experienced as a defeat that galvanized much of the younger aristocracy (the 'generation of 1812') into discovering or inventing the Russian nation and people. Again, there was a wave of reformism (including the first plans to emancipate the serfs) that resonated well beyond the unsuccessful Decembrist revolt led by reform-minded former officers in 1825. Tolstoy's *War and Peace* (1868–9), in which Pierre Bezukhov's resolve to stay in Moscow and assassinate Napoleon epitomizes the rejection of defeat, and Tchaikovsky's '1812 Overture' (commissioned for the 1882 Moscow exhibition) suggest that the touchstones of defeat and victory in 1812 remained important for Russia down to the Revolution.[3]

Perhaps the most important example in the twentieth century of a defeat that in this case was incompletely redeemed by subsequent victory is the fall of France in 1940. For the military defeat was envenomed by the responses of the Vichy regime, and of a spectrum of collaborators, to German occupation. These responses were only partly offset by internal Resistance, the Free French and eventual Liberation by the Allies. This all left such a contradictory legacy that the heroic post-war myths of the Gaullist and Communist Resistance broke down under the influence of a recurrent 'Vichy syndrome' (brilliantly analysed by Henri Rousso), through which the memory of defeat has haunted the French almost to the present.[4]

This first category, then, reminds us that temporary defeat may shake a state or nation to the core despite subsequent victory, or even because of it. Yet definitive defeat – the final verdict of a war that shapes the subsequent peace – is even more central to nineteenth- and twentieth-century history. In particular, there is strong continuity in the role played by defeat in the sequence of diplomatic confrontations, wars and shifts in the balance of power that characterized Europe from the Battle of Waterloo to the Second World War. The acceptance of war as an instrument of foreign policy meant acknowledging that defeat (like victory) was a pivotal moment in this evolution, one full of significance for future developments (including the emergence of nation states) and characterized by a certain common understanding between the opposing camps.

Although the wars in question – the Napoleonic Wars, the wars of Italian and German unification and the two world wars – became more intense and destructive, they were premised on the continued existence of the major states as autonomous agents capable of responding to defeat as well as to victory.

One might even speak of a ritual of defeat in these national wars. The victors signalled their status by parading through the enemy's capital (for example, the Germans on the Champs-Elysées in 1871 and 1940), the vanquished signed a peace treaty accepting the outcome of the war including an indemnity and the victors remained in temporary occupation to ensure the implementation of the peace terms. So established was the ritual that it furnished a cumulative symbolism of revenge, with Bismarck basing the French indemnity in 1871 on that levied by Napoleon on Prussia at the Treaty of Tilsit in 1807, the Allies imposing the Treaty of Versailles in the Hall of Mirrors where the German Reich had been declared in 1871 and Hitler retrieving from a museum the railway carriage in which the Armistice had been signed in November 1918 for the reverse process in June 1940.

But defeat was, of course, much more than a ritual. It proved as powerful a stimulus as victory to the creation and definition of national identities and nation states. Two of the most influential texts defining the nation in the nineteenth century were Fichte's 'Addresses to the German Nation', delivered at the University of Berlin in 1807–8, and Renan's lecture at the Sorbonne in 1882 on 'What Is a Nation?'. Their divergent views are well known, Fichte distilling the Romantic concern with German particularity into a political programme and Renan rejecting definitions of the nation in terms of religion, race or geography in favour of a collective will to live together (the famous 'daily plebiscite').[5] But both texts were a response to defeat. To put it in a different way, the conflict between revanchist nationalism and a more tempered liberal patriotism which took place in France after 1870 was a struggle to resolve the legacy of defeat, and it had immense consequences for French political culture, resulting in the predominance of a moderate Republican nationalism, which turned the loss of Alsace-Lorraine into cultural nostalgia. A struggle over defeat with a quite different outcome characterized Weimar Germany. But in both cases, the continued independence of the defeated nation and the freedom of its political system to address the issues posed by defeat were crucial.

In this regard, the Second World War marked a break with the prevailing tradition of defeat. For Germany fought the war in the east on the quite different premise that most enemy states would be annihilated and their populations subordinated to the Nazi racial order. Even in the West, traditional features of defeat (armistice, occupation, indemnity) were used for unprecedented economic exploitation, not to mention compliance with genocide. Aspects of this approach (such as the harsh economic treatment of civilians) had already been applied by Germany to peoples defeated during the First World War. But in the extreme form of the Nazi project of continental domination, this was defeat of a different order. Not surprisingly, it had a corresponding impact on the Allied view of victory. This was summed up in 1943 by Harold Nicolson, who, as a fledgling diplomat, had participated in the Paris Peace Conference in 1919. 'Total war implies total victory', Nicolson wrote, 'and thus when the moment of peace arrives there is complete strength on the side of the victors and complete weakness on the side of the vanquished.'[6] In this third category, total defeat, the enemy is stripped of political sovereignty until it has been remade in the image of the victor.

This might seem a particularly harsh form of defeat. It is true that the complete subjection of the enemy in a struggle without compromise, and by a technology allowing massive force to be deployed against the homeland, produced a final paroxysm in the war against Germany and Japan that was unparalleled in its violence. Defeat meant physical destruction as never before. It was also followed by a mass transfer of populations in central Europe in which a desperate effort was made to force peoples to fit borders in redefined and ethnically purged states – an ironic coda to the Nazi project. The expulsion of the Sudeten Germans in 1946, which had been planned by the democratic Czechoslovak government-in-exile, is a case in point.[7] The Allied powers also formulated drastic projects for demilitarizing and even deindustrializing Germany and Japan, plans that survived in the dismantling policies of the early post-war years (especially severe in the Soviet zone of Germany).

But a hallmark of total defeat was also the reconstruction of the vanquished according to the political values and economic systems of their victors, along with the co-option of native traditions (Communism in East Germany, Catholicism and Social Democracy in West Germany, the retention of the emperor in Japan). Nicolson observed that 'every peace settlement should be regarded, not as the consummation of a given victory, but as the foundation of a slow process of reconstruction.'[8] That is what happened in the context of the Cold War. Nonetheless, the reinvention of a national community required by total defeat generates powerful tensions with the prior realities that have foundered in the catastrophe.

By analogy, this suggests a fourth category – internal defeat in civil wars. For the winner in these conflicts, by definition, seeks to deprive the loser of autonomy and to rebuild unity in accordance with their own values. The tendency is thus to seek total defeat – the declared goal of Generals Sheridan and Sherman in the American Civil War.[9] Yet here, too, the type of war shapes the 'slow process of reconstruction' referred to by Nicolson. The 'Reconstruction' imposed on the South by the Union after the American Civil War notoriously failed, leaving the former confederate states free – once they accepted an end to slavery and secession – to make a new racial order and eventually to engage in reconciliation with the north through the patriotism generated by the Spanish–American War and the First World War.[10]

However, what proved possible in a democracy with many shared values (despite the high price paid in a legacy of racial conflict) was inconceivable in the ideologically polarized civil wars of revolutionary Russia and the Spanish Second Republic. Here, silent conformity and (especially in Russia) the reinvention of social and cultural identities as a mask of survival were the lot of the defeated – with renewed repression at any sign of dissidence.[11] The chance of revenge delivered by the Nazi invasion to non-Russian nationalities in the Soviet case simply intensified the repression once Stalin reasserted his control, though partisan warfare prolonged the moment of final defeat in some cases until the early 1950s.[12] In Spain, too, there was some low-level insurgency throughout the 1940s. But in the main, the only alternative to silence was exile. As with Russian émigrés after the Civil War, this froze defeat in an eternal re-enactment designed to reverse the fatal moment. Increasingly at variance with reality, it resulted in a world apart, which is vividly portrayed by Jorge Semprun in his screenplay for Alain Resnais's film on the world of the Spanish Republican émigrés, *La guerre est finie* (1963).[13] When the reversal came, with Franco's death in 1975 and the rise of Gorbachov in 1985, the link with the generation – and cause – of the defeated had gone.

One final category might be noted, partial defeat, whereby a state suffers a military and diplomatic setback without being threatened in its territorial or political integrity. Colonial defeat has often been of this kind – whether at the hands of the colonized or rival imperial powers. Initial setbacks in the South African War of 1899–1902 (as well as the ultimate cost of victory) prompted nagging doubts in Edwardian Britain about the burden of empire and even the vitality of the 'race'. This contributed not only to a vein of imperial pessimism but also to the diplomatic and military realignment that eventually brought Britain into the First World War. More fundamentally, defeat in the Spanish–American War, with the loss of Cuba and the Philippines, prompted the cultural crisis which defined the intellectual 'generation of 1898' in Spain and precipitated the political contest over modernization that lasted until the Civil War.[14] The threat of French defeat in Algeria and American defeat in Vietnam proved turning points in the conduct of foreign policy and, in the French case, in the form of regime. De Gaulle in effect liquidated an unwinnable war, investing the political credit in a presidential Republic and the pursuit of French *grandeur* in foreign policy, which together provided his ultimate answer to the defeat of 1940.[15] The rise of the neoconservative trend in American politics from the Reagan presidency to its dominant position under George W. Bush can be seen, among other things, as a response to defeat in Vietnam. As such, it had already helped secure American supremacy in the last phase of the Cold War before providing a new orientation for the post-Cold War world.[16] Thus partial defeat is usually less devastating than temporary defeat. It acts as a catalyst for the modification of domestic or foreign policy and of broader cultural horizons.

The significance of the time frame since the French Revolution

Temporary, definitive, total, internal, partial: this typology of defeat is approximate and certainly not exhaustive. But all these kinds of defeat could be found in earlier eras. What, if anything, is particular about the time frame since the late eighteenth century? Is it merely a manageable span of recent history, or does it have an internal coherence?

There are good reasons to suggest the latter. Some military and political trends since the the French Revolution redefined defeat as they did much else. Most obvious is the type and scale of involvement in warfare by the populations concerned. Once popular sovereignty became the legitimizing principle of politics, it allowed the figure of the citizen-soldier to emerge as the basis of military recruitment and also as the goal of the nation organized for war. First imagined by the *levée en masse* with which the Convention met the threat to revolutionary France in August 1793, total mobilization was pursued by different regimes down to the Second World War. Of course, the ideal was never achieved. But it resulted in a degree of popular involvement in the business of war (fighting, producing, nursing, propagandizing), which far exceeded that possible in earlier periods.[17] The ability to replicate this process became part of the transfer of Western models of development to other parts of the world, such as Ottoman Turkey and Meiji Japan. When French military observers looked at the battles of the Russo-Japanese War, what impressed them above all in the fighting ability of the Japanese troops and their ability to withstand high losses was the discovery of a patriotic motivation that had hitherto been associated only with armies in Europe.[18]

When defeat came after such episodes of national mobilization, the psychological shock was all the greater since it cancelled the collective enterprise in which much of society had been engaged, required self-examination by groups and individuals (or myths and mechanisms that avoided this) and, in the case of total defeat, necessitated the obliteration or reinvention of the past.

At the same time, the emergence of the nation state produced a different legal and affective relationship of people with territory. The ideological geography of nationhood (as conceived most famously by Mazzini) created a new vocabulary of national integrity and hence of irredentist yearning for those excluded from the national fold. Nostalgia for a nation that was often still unachieved was typically formulated by intellectuals and activists. However, from the mid-nineteenth century it exercised a growing influence on the peoples of aspirant and emergent nation states. It turned war and politics into metaphors for each other. A partial victory became a defeat for those who had failed to achieve their maximum goals, as in Italy for the Mazzinian left after 1861 and for D'Annunzio and Mussolini in 1919–20. Military defeat, recast as martyrdom, became an instrument of nationalist mobilization. This was something the Fenian movement in mid-nineteenth-century Ireland developed with particular sophistication.[19] But the phenomenon has acquired a new relevance in the context of 'asymmetric warfare' in the contemporary world, in which defeat at the hands of a conventional army (such as that of the US and UK in Iraq) may feed the motivation of martyrdom in guerrilla and terrorist reprisal, whose explicit goal is the creation of communitarian or national solidarity and resistance. Reversing defeats that alienated the nation from its land and people could be presented as a matter of collective life or death, as in Germany after 1918. This made Willy Brandt's acceptance in 1970 of the new Polish–German frontier a turning point in the normalization of defeat after the Second World War. For he was renouncing West German claims to territory that had been part of Prussia since the seventeenth century.[20]

The modern state provided a highly differentiated framework for processing defeat. Bureaucratic power resulted in new institutions that could reflect on downfall or setback, often from different perspectives, such as the university or the professional military. The general staff, in particular, provided an organization whose task was to analyse victory and defeat, though this did not exempt it from myth-making and illusion. At the same time, mass politics produced a civil authority that might take a very different line and, in the wake of defeat, assume charismatic forms (Boulanger in 1880s France, Mussolini and Hitler after the First World War). The political response to actual or presumed defeat in Germany and Italy in the interwar period emerged in a field polarized between the military on the one hand and dictators on the other who promised national redemption, so that when the latter resorted to war it was resisted by the military. This was very clear in the case of Italy, where the army imbibed fascist ideology less than the Wehrmacht did in Nazi Germany.[21] Defeat, in other words, may provoke different responses in various parts of the state apparatus so that its interpretation becomes a major source of institutional political conflict.

Finally, the concepts and language available for dealing with defeat were transformed in this same period. Because defeat confounds expectations, it challenges established truths far more than victory, which tends to confirm the intellectual basis on which it has been won. This explains the galvanizing effect of temporary, partial and definitive defeats on the societies that sustain them. But the terms in which this happens change. Rulers fighting wars in early modern Europe relied on human agency and calculated outcomes just like their successors. But they also attributed defeat (even in a just war) to the will of God. Religious explanations by no means disappeared in the nineteenth and twentieth centuries. The basilica of the Sacred Heart that dominates Paris was built to expiate the defeat of 1870 and was consecrated in celebration of victory in 1919. Nonetheless, the secularization of Western thought involved ascribing human causality to all historical developments, which made defeat a pivotal issue.

Sometimes, the explanations could be excessive. Historians have stressed the contingent aspects of certain historic reversals. James McPherson has suggested that the Confederacy matched the Union more closely than is usually supposed. Julian Jackson has argued that the French and British came close to winning in 1940. And no less a person than the Duke of Wellington thought Waterloo 'a damned nice thing, the nearest run thing you ever saw in your life'.[22] Yet once it happens, defeat becomes a bottleneck of historical causality, making almost everything relevant. The point is illustrated by two of the classics of defeat literature. Ernest Renan's reflections of 1871 called for nothing less than the 'intellectual and moral reform' of France, adapting the best of the alien Prussian model (such as education) to French needs. Marc Bloch's *Strange Defeat* (penned in the months following the debacle of 1940) proposed a root-and-branch renewal of the institutions and entire social groups on which he blamed the national collapse.[23]

Granted this tendency, not all ideologies deal with military defeat in the same way. Liberal-democratic thought has tended to allow a differentiated if overdetermined judgement on what has gone wrong and what has to change, as in the reform programmes which emerged in the French and other Resistance movements during the Second World War – to which Bloch contributed. By contrast, both Social Darwinism and Marxist socialism placed conflict at the heart of their theories, which gave war (and thus defeat) a fundamentally determinist character, though in very different ways. War being the natural, cultural or biological order for Social Darwinism, victory or defeat was the ultimate judgement of history. Decline leading to defeat thus became an inevitable mechanism of evolution, as in Oswald Spengler's *The Defeat of the West*, which was conceived in a mood of cultural pessimism before the First World War but published after it, and which anticipated a cataclysmic 'final battle between Democracy and Ceasarism'.[24] The extreme form in which Hitler applied such beliefs to foreign policy admitted of no appeal. His defiant insistence on self-annihilation was a vital factor in the total defeat of 1945, which had its parallel in the culture of the Japanese military.[25]

In the case of Soviet Marxism, however, military conflict was an adjunct to class war, which was seen as the sociological mainspring of history. At least in theory, war was a dependent variable of revolution. Of course, this did not prevent the USSR from waging war as a great power, but it affected the way it imposed defeat. Conformity to the new social order was the price of survival after the Civil War, and also after the Second World War in Eastern Europe. In the latter case, class theory supplied the myths that helped convert total defeat into a new Soviet empire. One of these myths equated the Soviet-imposed state with the proletarian revolution. Another presented defeat as liberation from the capitalist past which alone had been responsible for fascism and the war. East Germany, in particular, could start with a clean ideological slate.[26]

Again, the list is indicative, not exhaustive, and perhaps I have simply argued that defeats reflect the conditions of their time. But identifying those conditions is clearly important for understanding what defeat has meant in the last two centuries and how this might differ from earlier periods.

Defeat as experience and memory

This brings me to the relationship between defeat and memory. Defeat in all its variants has at least two dimensions – event and process. As an event, it is the moment when the roles of victor and vanquished are fixed. It is the decisive battle, the act of surrender, the victory parade, the peace treaty or the flight into exile. It is Appomattox courthouse in 1865, the railway carriage in 1918 and 1940, Emperor Hirohito's radio address on 15 August 1945 or the last helicopter out of Saigon. It is the moment that ruptures time, replacing historical continuity with a new and uncertain horizon for the defeated.

But defeat is also the process by which this new reality is grasped and evolves into acceptance, normalization or rejection. Depending on the kind of defeat, many things will happen in that process. The reasons for the defeat will be sought. Soldiers will return home. The dead will be buried, and their sacrifice usually commemorated. Peace terms may be signed. Individuals may be prosecuted for war crimes or genocide. The former enemy may impose an occupation. Economic reconstruction and the rebuilding of shattered cities will begin. Lessons will be drawn. Plans for the future will be laid. Reconciliation with the enemy will be posited. Revenge or reversal of the defeat may be plotted. And individuals will try to pick up the threads of their lives. When the process is over is hard to say, since it leaves a permanent legacy, but it is perhaps at the point when some new equilibrium, some new sense of normality, is reached. This might even be seen as a third phase – the place defeat occupies in the history of a society once normalization has occurred.

Of these three dimensions, the second – the process of defeat – is central. But with regard to memory, it is also the most problematic. For the moment of defeat is by definition the transition from past to present, and thus is not likely to be forgotten. And the history of a defeat is the question of how the whole episode is remembered in later periods by the society involved. But memory is only one function of the *process* of defeat, and it may not always be the dominant one. For example, an overwhelming imperative in dealing with the immediate aftermath of an internal defeat, such as the Spanish Civil War, or a total defeat, such as that of Germany and Japan in 1945, is survival in the present – finding a place in exile or in the ruins of bombed-out cities where life can continue. Planning for the future may also be more important than dealing with the past. In the case of partial defeat, drawing the lessons and moving on may be paramount, as it arguably was for Donald Rumsfeld, Richard Cheney and Paul Wolfowitz as they launched a new crusade to contain Soviet power in the first Reagan administration, and began to anticipate the post-Cold War world.[27] The same may be true of the temporary defeat suffered by Prussia in 1807 or the definitive defeat of France after 1870, when the impulse of reform and renewal turned on making things different from how they had been. Abolishing serfdom or introducing compulsory primary schooling, for example, were conceived with the long-term future in mind. The future is even more important in the case of total defeat, when both trauma and control by the enemy are greater. Contemporary observers were stunned by the speed and enthusiasm with which, at least superficially, the Japanese turned their backs on the war and militarized nationalism and embraced the democratic revolution imposed by the Americans from 1945 to 1951.[28] The economic miracles of Japan and West Germany in the 1950s and early 1960s firmly oriented both societies to the future as the framework in which daily life and a new normality could be constructed.

The process of defeat may best be approached as an experience. It is lived, like any experience, in a temporal universe bounded by present and future as well as past. The past is especially important because the moment of defeat, the point at which continuity has been disrupted, remains inescapable. Memory has the crucial function of dealing with the disjuncture of defeat – that is, with the difference between the world before defeat (which may already have been a disturbed or traumatic one) and the world of defeat. As in any experience, memory is an active, not a passive, function. It processes the past in the form of narratives, rituals and symbols. It may be afflicted by trauma (those things that cannot be processed) and by amnesia (those things that are forgotten or suppressed, whether involuntarily or deliberately). But the point is that memory is only one aspect of the process, albeit a vital one. The past has to be seen in relation to the present and the future.

This is especially important because what, with memory and hindsight, is adjudged a particular kind of defeat may appear differently to contemporaries. The case of France in 1940 provides an illustration. The subsequent view that 1940 falls into the category of temporary defeat is justified in terms of the history of France post-1944 or indeed of that of the Resistance and the Gaullist Free French during the war, which they insisted was not over. But much about the Vichy regime and its collaboration with the Germans makes full sense only when one understands that Pétain and those around him *believed* that defeat was definitive. In the midst of the maelstrom, they had moved into a period of post-war reconstruction in which they sought to rehabilitate France after its shattering defeat and to negotiate a margin of manoeuvre in Nazi-dominated Europe. The war kept breaking in and the terms of the Faustian pact (which was largely construed by the French in the face of German indifference) kept deteriorating, drawing Vichy into increasingly lethal compromises especially in terms of implementing the genocide of the European Jews. But for Vichy what counted was a miscalculated future built on the apparently definitive defeat of 1940. There have been other such cases where the measure of defeat at the time and the temporal universe built upon it have dissolved in the face of a different long-term verdict, though not before creating their own historical impact with lasting consequences.

That said, is it possible to identify any distinctive roles played by memory as such in the experience of defeat? I believe it is, and I shall suggest several. The first and the most obvious is that of processing the moment of defeat. For explaining the downfall extends to the war that produced it, and raises the question of who is responsible for the overall failure. The type of defeat is important here. A heroic defeat from which images of military glory can be rescued may help preserve the influence (or at least the reputation) of those responsible. This was arguably the case with the Confederacy, where the persona of General Lee redeemed the surrender at Appomattox by a largely mythic reputation for chivalry and military genius.[29] Likewise, the *levée en masse* declared by the government of Gambetta in the autumn of 1870, and the continued fighting in which it resulted, enabled the French subsequently to claim a moral victory from a military defeat. Even here, there may be a scapegoat for the actual collapse. In the French case, it was Bazaine, the last of Napoleon III's generals, who surrendered the imperial army at Metz.[30] But in both cases, the American South and the Third Republic, the idea of heroic resistance contributed to the myth that a superior civilization (which still subsisted) had been worsted by the brute force of an intrinsically barbaric enemy. In Republican France, this constituted one of the legitimating beliefs of the new regime.

By contrast, the ambiguities of German defeat in 1918, in which the army did not fight to the bitter end on German territory (despite the views of some soldiers and civilians that it should), plus the democratic revolution that accompanied it, left opposed views as to the nature and cause of the collapse. The innate tendency for the various parties to scapegoat each other, and in particular for the military to blame the politicians, assumed the well-known form of the 'stab-in-the-back legend' on the part of the former supreme commanders and the nationalist right. Combined with an unrelenting campaign to reject the Allied view of the *Kaiserreich's* responsibility for the outbreak of war and wartime atrocities, this resulted in a memory of the conflict for many in Germany which conflicted with that of the former enemies, undermined reconciliation and eventually fostered a new conflict.[31] The position was the reverse in Germany and Japan after 1945. For even if there was the same rush to blame others, plus somewhat arbitrary judicial proceedings (epitomized by the Allied refusal to try Emperor Hirohito as a class A war criminal for going to war), the fallen regimes were this time made comprehensively responsible by public opinion for both war and defeat – and not just because of political re-education programmes.

A second issue, however, considerably complicates the task of coming to terms with the responsibility for war and defeat – the dead and returning soldiers and servicemen. For it is here that both the rhetoric of wartime sacrifice and the human cost of the conflict for the society prosecuting it are most intense, and hence that the disjuncture between war and defeat is at its sharpest. The opposite is the case with victory, which redeems the military sacrifices of the living and the dead (and hence the suffering of their families) in a process that finds expression in national commemorative rituals. With defeat, the return of the defeated soldier is perhaps politically and morally easiest (though not necessarily materially so) where defeat has been temporary (the one and a half million French prisoners of war liberated in 1945) or where the defeated regime manages to distil a positive message from the lost war. In France after 1870, for example, the semi-official organization *Souvenir Français* erected monuments to the war dead on battlefields in France and eventually even in German Alsace-Lorraine.[32] In post-1918 Germany, too, the veterans were powerfully represented in organizations spanning the political spectrum. However, the divisions over the war and responsibility for the defeat were such that a national memorial and ritual equivalent to those in London and Paris (at the Cenotaph in Whitehall and the Arc de Triomphe) proved impossible.[33]

Even in the case of partial defeat, the veteran may be an awkward reminder of an earlier state of mind in a world that now thinks differently about the conflict, as with American Vietnam veterans in the decade following the war.[34] In the case of total defeat, however, the comprehensive repudiation of the war for which society had been massively mobilized only a few years before makes the returned serviceman particularly hard to absorb. Revelations about the brutality of Japanese military conduct abroad reinforced the veterans' status as pariahs at home in the immediate post-war period.[35] This plight is compounded by the humiliation of being occupied by the former enemy and also by the fact that prisoners of war (who may amount to millions of the defeated at the war's end) are typically used for labour by the victor for years after the conflict, often in harsh conditions. In both Japan and West Germany, a general pacifism and anti-militarism provided a commemorative lowest common denominator as the only way of making sense of the war dead and the survivors.[36]

At some point in the process, the question arises (without necessarily being posed) of whether defeat can furnish the basis of a new normality or whether it remains unacceptable. The answer, either way, represents a third issue that shapes how war is remembered and defeat is understood. For if defeat is rejected, this preserves or resurrects the culture of wartime. Thus, the 'stab-in-the-back' myth in Weimar Germany drew on a wartime vision of the internal enemy (the pacifist, Jew and socialist) and sought to reverse defeat even at the price of a new war. Conversely, accepting defeat implies reconciliation with the enemy. This entails a process of 'cultural demobilization' that supplies one path to the normalization of peace. It turns on a relativization or reversal of wartime truths, including the meaning of defeat.[37]

In the most extreme form, defeat is reinterpreted as liberation from the prior form of one's own society or regime, so that for all the difference in their power relationship to each other, victor and vanquished come to share core values. This characterized the remaking of the political cultures of Germany and Japan after the Second World War. In less extreme cases, reconciliation with the former enemy by a country that retains its autonomy may mean seeking concessions on the defeat itself while stressing common endeavours (peace, social reform and shared political values) that dissolve enmity and repudiate conflict. This was the tenor of those currents in Germany and the former Western Allies that sought reconciliation and a new diplomatic order in the second half of the 1920s, through the Locarno treaties and the League of Nations. As hostility was replaced with cooperation, the real enemy came to be seen as war itself and the sting of defeat was lessened.[38] The initiation of economic integration achieved this result triumphantly in Western Europe in the 1950s.

A fourth, related theme in the process of defeat is that of regeneration. Renewal covers a spectrum of values and activities, from legal codes to religious belief and the physical environment. But in each case, the issue of whether responding to defeat means building on the past or breaking with it is crucial. The point can be put briefly. Confronting the rupture of defeat forces a society to address the past in ways the victor can avoid – what I have called the 'galvanizing effect'. This may mean self-consciously drawing on the past (even if it is substantially invented), appealing to a new future or more often doing both simultaneously. The cult of the Revolution, with its litany of great dates and great men, was fundamental to the ways in which the Third Republic overcame defeat (and there was a Catholic martyrology culminating in the cult of Joan of Arc that achieved the same end). Yet neither Republicans nor Catholics dwelt in the past, but promoted reforms to rebuild France, including universal military service, legitimized by the revolutionary ideal of the citizen-soldier, which created the army of 1914. In even starker fashion, the American South after 'Reconstruction' developed a cult of the 'Lost Cause' while projecting plans (and fantasies) of renewal through the idea of the 'New South'.[39] High culture in Weimar Germany was notoriously polarized between the modern and an invented traditionalism.[40] The rebuilding of German cities after the Second World War evinced a mixture of modernist architecture (inherited from Bauhaus via American emigration) and the careful reinvention of historic districts such as medieval Munich. Regeneration necessitated a selective processing of the past. While not confined to the defeated, it was more charged for them than for the victors.

One final aspect of the process of defeat must be mentioned. This can best be summed up as the unbidden return of trauma. I am not sure whether by this we are talking about a collective psychological mechanism or a metaphor for a cultural process. But the tension between the suffering and dislocation caused by defeat (and the war that preceded it) and the coping tendencies involved in processing defeat are such that normalization will remain fragile and subject to breakdown over a more or less lengthy period – depending on the severity and totality of the original experience.

The issue has been most fully explored in Henry Rousso's work on the Vichy syndrome, to which I have already referred. In this case, the defeat of 1940 and the divisive experience of collaboration to which it gave rise were processed after the Liberation by the heroic myths of Resistance (Gaullist and Communist) which reached their apotheosis in the transfer of Jean Moulin's remains to the Pantheon in 1964 in a ceremonial designed by André Malraux to reassert the continuity of Republican values. Yet within a few years, the documentary film by Marcel Ophuls, *The Sorrow and the Pity*, and the book of the American historian Robert Paxton on *Vichy France* (1972) had shattered the fragile consensus (whose purpose had been to bury the past and turn to the future) by exposing the alternative and corrosive memories of French collaboration and participation in the Holocaust. All this amounted to a 'past that would not pass' (to borrow the title of another book by Rousso and the journalist, Eric Conan).[41]

But the case could be made for continental Europe more widely. The short cuts and limited trials (however necessary the latter may have been) which marked the immediate post-war period gave way to the conformity and political conservatism of the 1950s in communist and democratic variants. As in France, selective myth-narratives of Resistance were used elsewhere to contain profound and conflicting memories of occupation, humiliation and complicity.[42] But in the case of Germany, the task was double – to reconcile guilt for crimes for which new legal and moral terms had to be coined ('crimes against humanity', 'genocide') with a deep sense of victimhood arising from the destruction of German cities and the death of half a million civilians by Allied bombing. Added to this were the seven million military dead and the twelve million Germans who had fled or were expelled from former German territories – and who shared in the ambiguities of German responsibility for the war. There were only a few redeeming fragments of resistance from which to fashion a countervailing narrative in tune with post-war values. So it is scarcely surprising if public silence and selective official memories weighed more heavily in Germany even than elsewhere.[43]

Yet the normalized world of post-war Germany (especially in its western variant) could not contain the trauma of the defeat and the events that had led to it, and so became the target for a younger generation in the 1960s as more critical political cultures emerged to reinterrogate the past. The process is by no means over. The controversy aroused since the 1990s by the issue of Allied bombing of German cities during the war (which lifted the last taboo on a self-perception of German victimhood) and Gunter Grass's revelation in 2006 that his role as guardian of West Germany's conscience regarding the war was built on silence over a youthful engagement in the *Waffen SS* both indicate that the moral and psychological consequences of defeat still affect Germany.[44]

All this has coincided with the long and painful process by which the survivors of the Nazi genocide of European Jews and the families of the victims at last began to speak and, above all, to be heard in the final decades of the twentieth century. Not just for France or even for Germany, but for Europe as a whole this was a 'past that would not pass'. So too it proved for Israel, for as Tom Segev has argued, it was not until the Eichmann trial in 1961 that the full magnitude of the Holocaust was accepted. It was as if a prior defeat became a retrospective foundation stone of the Jewish state.[45]

This final point brings us full circle to the importance of the subject of defeat. For the very magnitude of the experience and its intrinsically traumatic nature make defeat something that challenges individuals and entire societies. Perhaps more than any other kind of episode, it opens people to the intellectual and political scrutiny of their own past while exposing them to the psychological consequences of what is a deeply threatening experience. The inescapable process of remembering has to confront both these dimensions of defeat while sustaining a present and building a future.

Notes

1. Wolfgang Schivelbusch, *The Culture of Defeat: On National Trauma, Mourning and Recovery* (2001; English translation, London: Granta, 2003).
2. Thomas Nipperdey, *Germany from Napoleon to Bismarck, 1800–1866* (1983; English translation, Dublin: Gill and Macmillan, 1996), pp. 1–84 ('The Great Upheaval').
3. Orlando Figes, *Natasha's Dance: A Cultural History of Russia* (London: Allen Lane, 2002), pp. 72–146; Leo Tolstoy, *War and Peace* (1868–9; English translation, London: Penguin, 1957, new edn, 1982), pp. 1066–1103; Didier Francfort, *Le Chant des nations: Musiques et cultures en Europe, 1870–1914* (Paris: Hachette, 2004), p. 134.
4. Henry Rousso, *The Vichy Syndrome: History and Memory in France since 1944* (1987; English translation, Cambridge, MA: Harvard University Press, 1991); Julian Jackson, *France. The Dark Years, 1940–1944* (Oxford: Oxford University Press, 2001).
5. Johann Gottlieb Fichte, *Addresses to the German Nation, 1807–8* (English translation, Chicago: Open Court Pub. Co., 1922); Ernest Renan, 'Qu'est-ce qu'une nation?' (1881) in Raoul Girardet (ed.), *Renan. Qu'est-ce qu'une Nation? et autres écrits politiques* (Paris: Imprimerie Nationale, 1996), pp. 223–43.
6. Harold Nicolson, *Peacemaking 1919* (London: 1933; revised edn, 1943, republished, Methuen, 1964), 'Introduction' (1943), p. xv.
7. Norman Naimark, *Fires of Hatred: Ethnic Cleansing in Twentieth Century Europe* (Cambridge, MA: Harvard University Press, 2001), pp. 108–38.
8. Nicolson, *Peacemaking*, 'Introduction', p. xvi.
9. Schivelbusch, *Culture of Defeat*, pp. 38–9, 191.
10. Eric Foner, *A Short History of Reconstruction 1863–1877* (New York: Harper and Row, 1990).
11. Sheila Fitzpatrick, *The Russian Revolution, 1917–1932* (Oxford: Oxford University Press, 1982), pp. 76–97; Sheila Fitzpatrick, *Everyday Stalinism: Ordinary Life in Extraordinary Times: Soviet Russia in the 1930s* (Oxford: Oxford University Press, 1999), pp. 115–38 ('Insulted and Injured'); Michael Richards, *A Time of Silence: Civil War and the Culture of Repression in Franco's Spain, 1936–1945* (Cambridge: Cambridge University Press, 1998), pp. 26–46.
12. Nicolas Werth, 'A State against Its People: Violence, Repression and Terror in the Soviet Union' in Stéphane Courtois, Nicolas Werth, Jean-Louis Panné, Andrzej Paczkowski, Karel Bartosek and Jean-Louis Margolin, *The Black Book of Communism: Crimes, Terror, Repression* (1997; English translation, Cambridge, MA: Harvard University Press, 1999), pp. 236–9; Jan Gross, *Revolution from Abroad: The Soviet Conquest of Poland's Western Ukraine and Western Belorussia* (1988; new edn, Princeton: Princeton University Press, 2002).
13. Jorge Semprun, *La Guerre est finie: scénario du film d'Alain Resnais* (Paris: Gallimard, 1966).

14. Donald Shaw, *The Generation of 1898 in Spain* (London: E. Benn, 1975), pp. 206–13; Raymond Carr, *Modern Spain, 1875–1980* (Oxford: Oxford University Press, 1980), pp. 47–70.
15. John F. V. Keiger, *France and the World since 1870* (London: Arnold, 2001), pp. 70–3.
16. James Mann, *The Rise of the Vulcans: The History of Bush's War Cabinet* (New York: Viking, 2004), pp. 52–5.
17. Daniel J. Moran and Arthur Waldron (eds), *The People in Arms: Military Myth and National Mobilization since the French Revolution* (Cambridge: Cambridge University Press, 2003); John R. Gillis (ed.), *The Militarization of the Western World* (New Brunswick and London: Rutgers University Press, 1989).
18. Olivier Cosson, 'Expériences de guerre au début du XXe (guerre des Boers, guerre de Mandchourie, guerres des Balkans)' in Stéphane Audoin-Rouzeau and Jean-Jacques Becker (eds), *Encyclopédie de la Grande Guerre: Histoire et culture* (Paris: Bayard, 2004), pp. 97–108.
19. Vincent R. Comerford, *The Fenians in Context: Irish Politics and Society, 1848–82* (1982; new edn, Dublin: Wolfhound, 1998), pp. 145–7.
20. Tony Judt, *Post-War: A History of Europe since 1945* (London: Heinemann, 2005), p. 498.
21. Macgregor Knox, *Common Destiny: Dictatorship, Foreign Policy and War in Fascist Italy and Nazi Germany* (Cambridge: Cambridge University Press, 2000), pp. 228–31; Giorgio Rochat, *Le Guerre italiane, 1935–1943* (Turin: Einaudi, 2005), pp. 169–205; Omer Bartov, *Hitler's Army: Soldiers, Nazis and War in the Third Reich* (New York: Oxford University Press, 1992), pp. 179–86.
22. James McPherson, 'American Victory, American Defeat' in Gabor S. Borritt (ed.), *Why the Confederacy Lost* (New York: Oxford University Press, 1992), pp. 17–42; Julian Jackson, *The Fall of France: The Nazi Invasion of 1940* (Oxford: Oxford University Press, 2004); Norman Davies, *Europe: A History* (Oxford: Oxford University Press, 1996), p. 762 (on Wellington).
23. Ernest Renan, *La Réforme intellectuelle et morale de la France* (1871; Paris: Union Générale d'Editions, 1967), pp. 144–64; Marc Bloch, *L'Etrange Défaite* (Paris: Editions du Franc-Tireur, 1946).
24. Oswald Spengler, *The Decline of the West* (1922; abridged English translation, 1959; new edn, Oxford: Oxford University Press, 1991), p. 397.
25. Gerhard L. Weinberg (ed.), *Hitler's Second Book: The Unpublished Sequel to Mein Kampf*, (1961; English translation, New York: Enigma, 2003), pp. 7–15; Ian Kershaw, *Hitler: 1936–1945. Nemesis* (London: Allen Lane, 2000), pp. 751–94; John Dower, *War without Mercy: Race and Power in the Pacific War* (London: Faber, 1986), pp. 262–90.
26. Jeffrey Herf, *Divided Memory: The Nazi Past in the Two Germanys* (Cambridge, MA: Harvard University Press, 1997), pp. 34–5.
27. Mann, *Rise of the Vulcans*, pp. 136–7.
28. John Dower, *Embracing Defeat: Japan in the Wake of World War II* (London: Allen Lane, 1999), pp. 26–7.
29. Schivelbusch, *Culture of Defeat*, pp. 64–7.
30. Ibid., pp. 128–59; François Roth, *La Guerre de 70* (Paris: Fayard, 1990), pp. 680–708.
31. Schivelbusch, *Culture of Defeat*, pp. 189–288.
32. Roth, *La Guerre de 70*, pp. 691–8.

33. George Mosse, *Fallen Soldiers: Reshaping the Memory of the World Wars* (New York: Oxford University Press, 1990), pp. 70–106.
34. Susan Jeffords, *The Remasculinization of America: Gender and the Vietnam War* (Bloomington: Indiana University Press, 1989).
35. Dower, *Embracing Defeat*, pp. 58–61.
36. For a reflection on this process in both countries, and its prolongation in Japan, see Ian Buruma, *Wages of Guilt: Memories of War in Germany and Japan* (London: Jonathan Cape, 1994).
37. John Horne (ed.), 'Démobilisations culturelles après la Grande Guerre', *14–18 aujourd'hui-Heute-Today*, vol. 5 (Paris, 2002).
38. John Horne, 'The European Moment between the Two World Wars, 1924–1933' in Madelon de Keizer and Sophie Tates (eds), *Moderniteit: Modernisme en Massacultuur in Nederland, 1914–1940* (Amsterdam: Nederlands Instituut voor Oorlogsdocumentatie, Waelbeurg Pres, 2005), pp. 223–40.
39. David D. Blight, *Race and Reunion: The Civil War in American Memory* (Cambridge, MA: Harvard University Press, 2001), pp. 255–99.
40. Peter Gay, *Weimar Culture: The Outsider as Insider* (1969; London: Penguin, 1974), esp. pp. 48–72 ('The Secret Germany'); Jeffrey Herf, *Reactionary Modernism: Technology, Culture and Politics in Weimar Germany and the Third Reich* (Cambridge: Cambridge University Press, 1986).
41. Henry Rousso and Eric Conan, *Vichy: An Ever-Present Past* (1996; English translation, Hanover: University of New England Press, 1998).
42. Pieter Lagrou, *The Legacy of Nazi Occupation: Patriotic Memory and National Recovery in Western Europe, 1945–1965* (Cambridge: Cambridge University Press, 2000).
43. Sabine Behrenbeck, 'Between Pain and Silence: Remembering the Victims of Violence in Germany after 1949' in Richard Bessel and Dirk Schumann (eds), *Life after Death: Approaches to a Cultural and Social History of Europe in the 1940s and 1950s* (Cambridge: Cambridge University Press, 2003), pp. 37–64.
44. Günter Grass, *Peeling the Onion* (2006: English translation, London: Harvill Secker, 2007).
45. Tom Segev, *The Seventh Million: The Israelis and the Holocaust* (New York: Henry Holt, 1991), pp. 323–84 (Eichmann trial) and 421–507 (memory).

3

Defeat and Foreign Rule as a Narrative of National Rebirth – The German Memory of the Napoleonic Period in the Nineteenth and Early Twentieth Centuries

Christian Koller

The Napoleonic period has played a crucial role in Germany's cultural memory[1] since the end of the anti-Napoleonic Wars. Between 1795 and 1805, Prussia had preserved its neutrality in the succeeding coalition wars. In 1806, it entered war against France, and on the 14 October its army experienced a disastrous defeat in the double battle of Jena and Auerstedt. About 20,000 Prussian and Saxon soldiers were killed or wounded, 13,000 were captured. The double battle proved that the Prussian army was outdated, poorly trained and inflexibly led. On the 27 October, Napoleon and his troops entered Berlin, while the Prussian king and his family fled eastwards. This disaster caused an enormous shock.[2] In the treaty of Tilsit in July 1807, Prussia remained an autonomous state, but it lost half of its territories. It was occupied by French troops and charged with heavy contributions. In the following years, the leading ministers Stein and Hardenberg enacted a wide-ranging modernization programme that included reforms of government, administration, agriculture, trade, taxation, military and education.[3]

Prussia's military defeat in the battles of Jena and Auerstedt and the following period of *Fremdherrschaft* (foreign rule) have always been perceived as a low point in German national history and which stood in sharp contrast to the 1813 uprising that climaxed in the so-called *Völkerschlacht* (peoples' battle) of Leipzig.[4] My contribution will analyse both this memory's narrative structure and its functions in the pre-unification period, in the Wilhelminian Empire, in the Weimar republic and in the Nazi era. I will argue that defeat and foreign rule became in Germany's political culture[5] a myth of national rebirth that linked the requirements of a foundation myth[6] to the notion of the very old age of the German nation. As the political scientist Rudolf Speth has shown, political myths are, according to a semiotic understanding of culture, an important element within the process of the construction of political reality, because they supply the cultural memory of political groups with sense.[7] In Herfried Münkler's words: 'Political myths ensure communities that what happened had had to happen, that events were not accidental, but necessary, and that they were more than mere events, that they owned an eschatological dimension.'[8] Thus, they play a crucial role within the narration of nationality that 'tells people a story about themselves in order to give sense to the social reality'.[9]

Pre-unification period

Between the end of the anti-Napoleonic Wars and the early 1840s, there was a significant shift in the perspective under which the Napoleonic period was seen. First, the cultural memory focussed on Napoleon as a tyrant trying to establish a `universal monarchy'. The concept of 'universal monarchy', having existed since the late Middle Ages,[10] was clearly pre-nationalist. Its opposition was not the liberty of the nations, but the balance of powers.[11] Historians such as Carl Venturini, who designated Napoleon as *Welttyrann* (world tyrant),[12] or Arnold Hermann Ludwig Heeren[13] interpreted the history of the Napoleonic era this way. Thus, they stressed the cooperation of European princes and peoples in the struggle against the Emperor of the French.[14] In the 1840s, this way of interpretation nearly vanished. The new paradigm was the dialectics between foreign rule and national liberty. Historiography and cultural memory now concentrated on the nation's struggle against the foreign oppressors (the so-called *Erhebung*).

Parallel to this shift of perspective, a new German term emerged, which had not existed during the Napoleonic age, namely *Fremdherrschaft* (foreign rule).[15] At first, it was exclusively used in speaking of the Napoleonic rule over Germany. From 1840 on, it became an abstract concept for interpreting the political world of past and present, for instance the Austrian rule in Italy[16] and the English rule over Ireland[17] or the Roman rule over Celts, Carthaginians or Teutons[18] and the Popish influence in Germany before the reformation.[19] Thus, the experience of the Napoleonic hegemony and the memory thereupon created a new political concept that had great influence on both the interpretation of the actual world and the memory of crucial events in Germany's national history.[20] Both, the shift in the memory of the Napoleonic era and the diffusion of the concept of *Fremdherrschaft* were indicators of an increasing nationalization of Germany's political culture.

The shift from the perspective of universal monarchy to the focus on foreign rule also changed the framework of emplotment. The phenomenon of Napoleon was no longer mainly a problem of world history, but of German national history. In this context, the defeat of Jena and Auerstedt and the following years of Napoleonic foreign rule were perceived as a low point in German national history, yet the overwhelming majority of these interpretations also saw it as a turning point. So, it was the most terrible evil, but in the long run, it nonetheless had positive effects.

The most common metaphors to describe these effects were *Wiedergeburt* (rebirth),[21] *Erwachen* (awakening)[22] and *Lehrjahre* (years of apprenticeship).[23] Roderich Benedix, for instance, stated in his *Volksbuch* (people's book) about the years from 1813 to 1815, published in 1842, that every national history had its climax that was followed by decline. German history's climax had been 600 years ago. From then on it declined until the Napoleonic period, which saw the great shame of French usurpers becoming German princes. Benedix further stated that every nation's life, like a man's life, runs through youth, manhood, old age and death. But a nation could rise from death. According to this analogy, the German people had run through a whole life, was very old in the age of the French Revolution and died in the wars until 1809. Yet in 1813, it rose and started a new life. Since then, it had moved towards a new climax, which would be reached with the political unity of the German nation. The 1813 uprising was a divine hint to indicate Germany's future direction.[24]

Benedix was not the only one to interpret the Napoleonic period in a religious manner. In 1841, Theodor Rohmer in his book *Deutschlands Beruf in der Gegenwart und Zukunft* (Germany's task in present and future) compared the Napoleonic foreign rule to Persian and Babylonian captivities of the Jewish people. Rohmer stated that God had sent Napoleon in order to purify the German nation, the second chosen people, that now had to establish a new world order.[25] In 1863, during the celebrations of the fiftieth anniversary of the *Völkerschlacht* of Leipzig, the Lutheran pastor Friedrich Ahlfeld gave a sermon comparing the history of the Napoleonic period with the biblical story of Samson. Both Samson and the German nation had left the path of virtue and God had punished both of them. The German nation's sin had consisted of the adoption of French ideas and manners in the age of Enlightenment and of certain sympathies for the French Revolution.[26] Another pastor, B. B. Brückner, paralleled the Napoleonic foreign rule with the Egyptian captivity of the Jewish people. Both Ahlfeld and Brückner thought that the wars of liberation had been part of the German nation's turn back to

God. Brückner, like Rohmer, was even sure that God had chosen the German nation to become the whole world's fecundating soul.[27]

Other interpretations lacked these religious connotations, but the structure of their argument was very similar. In May 1848, the newspaper *Deutsche Allgemeine Zeitung* stated that the Napoleonic foreign rule had been a *Sühnung* (expiation) and a *Bluttaufe* (bloody baptism).[28] In 1863, Leipzig's Lord Mayor Dr Koch meant that the foreign rule as a *Zeit der Prüfung* (time of examination) had been necessary to redeem the sins of the preceding centuries.[29]

Thus, both the religious and the secular-teleological interpretations of defeat and foreign rule followed a classical structure of myth: paradise – fall – purification – rebirth. The 1806 defeat marked the transition from fall to purification. Conservatives and liberals shared this narrative structure. These two main political tendencies only differed in the interpretation of the years after 1815. Whereas the conservatives thought that Germany had gained new national grandeur in the wars against Napoleon, the liberals claimed that there was still a long way to go until Germany's national history would have reached a new climax. For the latter, then, the myth of national rebirth was a legacy and an urge to action, not yet a matter of triumph.

The only dissident voices emerged from the radical left. Already in 1819, Karl Follen, an exiled radical refugee in France, criticized the *tyrannie fraternelle et paternelle* now reigning in Germany which was much worse than the former *tyrannie étrangère*.[30] Ludwig Börne stated in 1832 that the war 'they called war of liberation' had liberated no one but the German princes.[31] In July 1848, Karl Marx stressed in an article in the democratic newspaper *Neue Rheinische Zeitung* that Napoleon had abolished the old feudal charges in the Rhineland. He even put the terms '*Fremdherrschaft*' and '*korsischer Tyrann*' (Corsican tyrant) into quotation marks.[32] Marx and Engels repeatedly designated pre-Napoleonic Germany as an Augean stable that had been cleaned by the emperor of the French.[33] Thus, radical interpretations tended to see Napoleon as a representative of the French Revolution rather than as a foreign oppressor. The Prussian defeat at Jena and Auerstedt appeared to them as a victory over feudalism, and the end of Napoleonic hegemony over Germany, rather than being a liberation, was a reactionary backlash.

Wilhelminian empire

After the German unification of 1871, the mainstream interpretation of the Napoleonic period remained unchanged. Defeat and foreign rule were still designated as a *Wendepunkt* (turning point),[34] *Läuterung* (purification)[35] or *innere Gesundung* (inner recovery);[36] the German reaction thereupon was a *Wiedergeburt* (rebirth)[37] or an *Erwachen* (awakening).[38] The liberal historian Friedrich Meinecke wrote in 1908 that under the winter cover of foreign rule, new green crops had grown.[39] And Heinrich Ulmann stated in 1914 that, like the Nile's wave used to flood Egypt disastrously but left behind fertilizing mud, the flood of the Napoleonic rule had changed the German fatherland.[40] During the celebrations of the hundredth anniversary of the *Völkerschlacht* of Leipzig in 1913, Emperor William II stated that the years between 1806 and 1813 had been a hard punishment for the preceding times of stagnation and decline.[41] The Catholic newspaper *Germania* wrote in the same context that the late eighteenth century's enlightened *Zeitgeist* had experienced a bitter defeat in 1806 and that the liberation of 1813 had been the work of a religious and national spirit.[42]

The unification of 1871 was generally perceived as a new climax in German national history, the fulfilment of the process of national rebirth that began in 1806. This view was shared by conservatives, liberals and Catholics, whose interpretations only differed slightly. Conservatives and Catholics tended to stress the ideas of Enlightenment and of the French Revolution as the main causes of the 1806 defeat, while liberals pointed at the feudal structure and the absolutist governments of pre-Napoleonic Germany. The victory in the anti-Napoleonic wars was in conservative eyes mainly to the Prussian king's credit, whereas liberals stressed the people's role and Catholics emphasized the religious spirit. Thus, all three of the political tendencies that shared in the Wilhelminian Empire's heterogeneous cartel of power accommodated the history of defeat, foreign rule and national rebirth to the needs of their political philosophy without changing the basic structure of its narrative.

This narrative also played an important role in historiography. The leading school of German historiography in the second half of the nineteenth century, the Borussian branch of historicism,[43] headed by Johann Gustav Droysen, Heinrich von Treitschke and Heinrich von Sybel, fitted the story of defeat and national rebirth into a dualistic concept. All political communities of past and present were either nation states or *Fremdherrschaften*. In this perspective, German history since the late seventeenth century was the continual struggle of Prussia, as the agent of German national unification, against several forms of foreign rule. The latter could be 'overt', as exerted by Napoleon, or 'covert', as in the emperorship of the Catholic and allegedly non-national Habsburgians of the Holy Roman Empire and their presidency of the German confederation between 1815 and 1866.[44]

However, there were at least two dissident ways of interpretation. For social democratic thinkers as August Bebel,[45] Franz Mehring and Kurt Eisner, following the tradition of pre-unification times radical interpretations, Napoleonic foreign rule had not been a low point in Germany's national history, but a substitute for the absence of a bourgeois revolution and as such it had been 'historical progress'.[46] However it could only be an incomplete substitute, because it did not overthrow the Prussian aristocracy. This, according to Mehring, had been a curse upon German history ever since.[47] For Eisner, the dissolution of the Holy Roman Empire and the founding of the kingdom of Westphalia had been the basic unit of a new modern Germany, but Napoleon's defeat had interrupted this development. So, the wars of liberation had in reality been wars to destroy liberty (*Freiheitsvernichtungskriege*).[48] In 1913, the socialist newspaper *Vorwärts* stated that the junkers had been overthrown in 1806, but had been heaved back into power in 1813.[49] During the celebrations of the hundredth anniversary of the *Völkerschlacht* of Leipzig, the Social Democratic Party held meetings entitled *Völkerschlacht und Völkertrug* (peoples' battle and peoples' fraud).[50]

Another dissident interpretation came from the adherents of political anti-Semitism, which became an organized political movement around 1880.[51] Wilhelm Marr, an important pamphleteer of the secular wing of political anti-Semitism,[52] stated in his best-seller *Der Sieg des Judenthums über das Germanenthum* (The Victory of Jewry over the Teutons) that at present, there was a foreign rule much worse than the Napoleonic one: the Jewish foreign rule. Whereas Napoleon had not succeeded in co-opting the German elite, the Jews had taken control of most of the political parties, large sectors of the German economy and especially over the press.[53] Marr was not the only anti-Semite to denounce an alleged Jewish foreign rule. The *Allgemeine Evangelisch-Lutherische Kirchenzeitung*, the leading newspaper of conservative Protestantism, published several articles entitled *Die jüdische Fremdherrschaft* (The Jewish Foreign Rule) around 1880/81.[54] And Adolf Stöcker, court chaplain in Berlin and founder of the anti-Semite *Christlich-Soziale Arbeiterpartei*, asserted in a Reichstag debate in 1892 that the anti-Semite movement had emerged because the German nation, after having returned from the unification wars, had become aware of the shameful foreign rule lasting upon it.[55] Thus, the interpretation of the present as a time of even worse foreign rule than during the Napoleonic era was shared by secular as well as religious anti-Semites.

Although the majority of the political elite did not share this mode of interpretation, its anti-Semite assumptions were widespread. According to Shulamit Volkov, anti-Semitism even became a 'cultural code' within the Wilhelminian Empire's conservative and nationalist elite, a token of an ideology subsuming anti-modernism, nationalism, cultural pessimism and imperialism.[56] So, it is small wonder that the basic structure of the anti-Semites' interpretation of defeat and foreign rule, its pessimistic assumption of a repetition of foreign rule and its notion of 'inner' foreign rule, became influential after the trauma of the defeat in the First World War.

Weimar Republic

After the First World War, the old mainstream interpretation of the Napoleonic foreign rule as the initial period of a national ascent climaxing in the present had become dysfunctional. However, the memory of 1806 continued to have an important function.[57] It now had to console German nationalists that as there had been Leipzig after Jena, there would be a new liberation after Versailles.[58]

Max von Szczepanski stated in 1922 in the conservative periodical *Grenzboten* that the defeat of 1918 was much more shameful than the one of 1806 because it had been initiated by a treacherous revolt from the ranks of Germany's own *Volksgenossen* (national comrades). As the 1918 sin had been greater than the one of 1806, the punishment would be harsher, too, and there would be a very long way to a new Leipzig. All the same, the history of the Napoleonic period showed the way the path towards liberation had to start: before the victory over the enemy, there had to be a victory over one's self.

Several nationalists as, for instance, Max von Szczepanski in 1922[59] or Adolf Hitler in his unpublished second book in 1928[60] stressed the role of outstanding leaders in overcoming the Napoleonic foreign rule and concluded that the Weimar Republic should be replaced by an authoritarian system. In March 1923, during the French and Belgian occupation of the Ruhr Area, Wolfgang Eisenhart stated in the conservative newspaper *Neue Preussische Zeitung* that the actual situation was a repetition of the beginning of the nineteenth century. Like the enlightened eighteenth century's cosmopolitanism had been replaced by nationalism due to Napoleon's hegemony over Germany, socialist internationalism and democratic republicanism would now be wiped out by the German reaction to the new foreign rule climaxing in the occupation of the Ruhr Area. Once again, the French were fated to be the catalysts of German nationalism.[61]

While the 'traditional' conservatives had the notion of a simple repetition of Napoleonic times, other right-wing forces, which were later subsumed under the term 'national revolution',[62] developed an eschatological perspective. They did not want a renaissance of the old Wilhelminian Empire after the 'foreign rule' of the Versailles treaty and the liberal and democratic Weimar order would be overcome, but something new, which the cultural historian and political writer Arthur Moeller van den Bruck was already calling a 'Third Reich' in 1923. Moeller van den Bruck, being close to the ideas of the 'national Bolsheviks' who propagated an anti-Western alliance between extreme right and extreme left, criticized the conservatives' notion of a new 1813 as reactionary. The new liberation would not follow the patterns of the anti-Napoleonic Wars, but it would be spearheaded by the working class. Its result would not be the empire the reactionaries dreamt of, but the empire 'of all of us'.[63]

The Nazi era

After the Nazi seizure of power, the parallels between the present and the time around 1800 remained an important element of Germany's memorial culture.[64] In addition to the parallel between 1806 and 1918, Nazi propaganda drew a parallel between 1813 and 1933. When universal compulsory military service was reintroduced in March 1935, the newspaper *Germania* wrote: 'Tilsit was followed by the uprising of 1813, Versailles is followed by the 16 March 1935!'[65] However, for Nazi propaganda, the 1813 uprising was only an unripe forerunner of 1933. Hans Erich Feine, professor of law in Tübingen, wrote in 1936 that during the decade of foreign rule, large parts of the German people had experienced a national spirit for the first time that had been a presentiment of the vehement experience of his days.[66]

The parallels between the Napoleonic foreign rule and the situation of Weimar Germany under the treaty of Versailles permitted the Nazis to legitimate almost every part of their policies as a step in the struggle against foreign rule. Already during the years of the Weimar Republic, the concept of foreign rule had consisted of two elements in the political language of the Nazis. First, there was an 'outer' foreign rule of the victors of the World War, directly exerted over the annexed and occupied territories, but indirectly over the whole German nation by means of the peace treaty restrictions.[67] The Nazis shared this notion with nearly all revisionists in Germany. Second, there was an 'inner' foreign rule, exerted by Marxists, democrats and especially by the Jews.[68] Within the logic of this pattern, the abolition of democracy as well as the persecution of Jews, socialists and communists were measures against inner foreign rule, and the territorial acquisitions (Saarland, Austria, Sudetenland) were the liberation of German brethren from the Versailles treaty's foreign rule.[69]

Arguing against the parallels between the Napoleonic period and the present was a very hard task for the illegal and exiled opposition as long as Hitler's foreign policy seemed to be successful. In 1937, a socialist pamphlet entitled *Von deutscher Freiheit* (On German Liberty) stated that the German nation had always been betrayed by its own elite. The defeat against Napoleon might be called shameful, but the shame was that a foreign conqueror had had to invade Germany to initiate modest reforms. So, the German people had to thank its reactionary elite for this shame. In 1813, the German nation had struggled not only against the foreign oppressors but also against the medieval barbarity of the German princes and for political unification and had not known that it would only exchange foreign rule against a stronger servitude under its own princes. A similar process was happening at present.[70]

The former Nazi Hermann Rauschning stated in 1938 in his book *Die Revolution des Nihilismus* (The Revolution of Nihilism) that the Nazi regime was a brutal form of inner foreign rule. All troubles of the Napoleonic era, which had still been vivid in the family's memory of the older generation, were now experienced once again in Nazi Germany.[71] Austrian socialists and communists again and again denounced the 1938 *Anschluss* as subjugation of the Austrian nation under Prussian foreign rule.[72] The use of the memory of the Napoleonic period by the Nazi propaganda thus forced Hitler's enemies to argue within the same narrative patterns – without much success.

At the end of the 1930s, when the German Reich started to invade territories inhabited by non-German-speaking populations, this sort of propaganda stopped suddenly. Only in the last months of the Second World War, as the allied forces progressed on German soil, the memory of 1806 regained its propagandistic function again. German newspapers shouted slogans like *Ein stolzes Volk geht nicht widerstandslos unter das Joch fremder Gewalthaber* (A proud nation does not go under the yoke of foreign tyrants without resistance), comparing at least implicitly the present situation with the years around 1806.[73]

Furthermore the last film production of the Third Reich was dedicated to this topic. Veit Harlan's movie *Kolberg*, ordered by Goebbels in summer 1943 after the Stalingrad disaster, told the story of the city of Kolberg (today's Kolobrzeg in Poland) after the defeat at Jena and Auerstedt. Under the command of Generalfeldmarschall Neidhardt von Gneisenau, Kolberg was the only Prussian town to resist the French troops until July 1807. The movie's premiere took place on the 30 January 1945 in the surrounded Atlantic fortress of La Rochelle, where the film spools had been dropped by parachute. The narrative of the 1806 defeat no longer emphasized the possibility of national rebirth, but urged heroic death; in the words of Kolberg's burgomaster Joachim Nettelbeck (played by Heinrich George): '*Lieber unter Trümmern begraben, als kapitulieren!*' (Better buried under ruins than surrender!).

Conclusion

The basic narrative structure about defeat and foreign rule remained unchanged during the whole of the nineteenth and early twentieth centuries. It followed a classical structure of myth: Paradise – fall – purification – rebirth. The 1806 defeat and the period of foreign rule were seen as punishment for non-national behaviour in the preceding decades and at the same time as a purification that prepared the German nation for its rebirth. However, there was a major change in this narrative after Germany's defeat in the First World War. Nineteenth-century memory had considered the story of fall, purification and national rebirth to be unique. After the First World War, this view was replaced by the notion of repetition: 1918 was a new 1806, Versailles was a new Tilsit and the Weimar Republic was a new *Fremdherrschaft*.

The functions of the memory of defeat and foreign rule changed several times. In the restoration period, the conservatives used the memory to stress the necessity of unity between princes and subjects, whereas the liberals emphasized the need for a unified national state. After the unification, both conservatives and liberals considered the defeat of 1806 as the initial event in Germany's national rebirth which climaxed in 1871. Thus, the memory of defeat and foreign rule attained the function of a foundation myth. The defeat of 1918 changed the function of the memory of 1806 again. It now had to console the nationalists and to give them hope that one day there would be a new Leipzig. After Hitler's seizure of power, these parallels were used to legitimate several elements of Nazi politics.

On balance, the memory of the Napoleonic period was flexible enough to accommodate changing political situations. Its basic structure became dysfunctional only after the Second World War, when it became clear that another repetition of the wars of liberation was neither likely nor desirable and German national myths had lost their legitimacy. In 1946, Friedrich Meinecke discussed in his famous book *Die deutsche Katastrophe* (The German Catastrophe) the question whether it was shameful to cooperate with the victors of the Second World War.

He argued that it was not, because the present foreign rule by the victors had been preceded by an inner foreign rule of the Nazis that had been much worse. A foreign rule from outside could purify a nation's soul, whereas an inner foreign rule caused much more damage.[74] Thus, the Nazi regime's crimes and the totality of the German defeat in the Second World War forbade any comparison between 1806 and 1945.

Notes

1. On the concept of cultural memory, see Jan Assmann, 'Collective Memory and Cultural Identity', *New German Critique* 65 (1995), 125–33.
2. See, for instance, Wolfgang Schivelbusch, *Die Kultur der Niederlage: Der amerikanische Süden 1865 – Frankreich 1871 – Deutschland 1918* (Berlin: 2001), p. 18.
3. See, for instance, Reinhart Koselleck, *Preussen zwischen Reform und Revolution: Allgemeines Landrecht, Verwaltung und soziale Bewegung von 1791 bis 1848* (Stuttgart: 1967).
4. See on the anti-Napoleonic Wars, for instance, Karen Hagemann, 'Of "Manly Valor" and "German Honor": Nation, War, and Masculinity in the Age of the Prussian Uprising against Napoleon', *Central European History* 30 (1997), 187–220; Hagemann, *'Mannlicher Muth und Teutsche Ehre': Nation, Militär und Geschlecht in der Zeit der Antinapoleonischen Kriege Preussens* (Paderborn: 2002). On the terminological debates, whether they were *Freiheitskriege* (liberty wars) or *Befreiungskriege* (wars of liberation), see Wolfgang Stammler, '"Freiheitskrieg" oder "Befreiungskrieg"?', *Zeitschrift für deutsche Philologie* 59 (1934), 203–8; Werner Conze et al., 'Freiheit' in Werner Conze, Otto Brunner and Reinhart Koselleck (eds), *Geschichtliche Grundbegriffe: Historisches Lexikon zur politisch-sozialen Sprache in Deutschland*, vol. 2 (Stuttgart: 1975), pp. 425–542, here: 504–5; Christian Koller, *Fremdherrschaft: Ein politischer Kampfbegriff im Zeitalter des Nationalismus* (Frankfurt/M–New York: 2005), pp. 212–15. On the memory of the *Völkerschlacht*, see Stefan-Ludwig Hoffmann, 'Mythos und Geschichte: Leipziger Gedenkfeiern der Völkerschlacht im 19. und frühen 20. Jahrhundert' in Etienne François, Hannes Siegrist and Jakob Vogel (eds), *Nation und Emotion: Deutschland und Frankreich im Vergleich, 19. und 20. Jahrhundert* (Goettingen: 1995), pp. 111–32; Kirstin Ann Schäfer, 'Die Völkerschlacht' in Etienne François and Hagen Schulze (eds), *Deutsche Erinnerungsorte*, vol. 2 (Munich: 2002), pp. 187–202.
5. On the concept of 'political culture', see Karl Rohe, 'Politische Kultur und ihre Analyse: Probleme und Perspektiven in der politischen Kulturforschung', *Historische Zeitschrift* 250 (1990), 321–46.
6. On foundation myths, see Wolfgang Kaschuba, 'The Emergence and Transformation of Foundation Myths' in Bo Stråth (ed.), *Myth and Memory in the Construction of Community: Historical Pattern in Europe and Beyond* (Berne: 2000), pp. 217–26.
7. Rudolf Speth, *Nation und Revolution: Politische Mythen im 19. Jahrhundert* (Opladen: 2000).
8. Herfried Münkler, 'Politische Mythen und nationale Identität: Vorüberlegungen zu einer Theorie politischer Mythen' in Wolfgang Frindte and Harald Pätzold

(eds), *Mythen der Deutschen: Deutsche Befindlichkeiten zwischen Geschichten und Geschichte* (Opladen: 1994), pp. 21–7.

9. Uri Ram, 'Narration, Erziehung und die Erfindung des jüdischen Nationalismus: Ben-Zion Dinur und seine Zeit', *Österreichische Zeitschrift für Geschichtswissenschaften* 5 (1994), 151–77, here: 153. See also Homi K. Bhabha, 'Introduction: Narrating the Nation' in Bhabha (ed.), *Nation and Narration* (London: New York: 1990), pp. 1–7.
10. See Franz Bosbach, *Monarchia Universalis: Ein politischer Leitbegriff der frühen Neuzeit* (Goettingen: 1988).
11. See Hans Fenske, 'Gleichgewicht, Balance' in Otto Brunner, Werner Conze and Reinhart Koselleck (eds), *Geschichtliche Grundbegriffe: Historisches Lexikon zur politisch-sozialen Sprache in Deutschland*, vol. 2 (Stuttgart: 1975), pp. 959–96.
12. Carl Venturini, *Russlands und Deutschlands Befreiungskriege von der Franzosen-Herrschaft unter Napoleon Buonaparte in den Jahren 1812–1815*, vol. 4 (Leipzig: 1819), p. I.
13. A.[rnold] H.[errmann] L.[udwig] Heeren, *Handbuch der Geschichte des Europäischen Staatensystems und seiner Colonien, von seiner Bildung seit der Entdeckung beider Indien bis zu seiner Wiederherstellung nach dem Fall des Französischen Kaiserthrons und der Freiwerdung von Amerika*, vol. 2, 5th edn (Goettingen: 1830), p. 269.
14. On Napoleon's image in Germany's memorial culture, see Hagen Schulze, 'Napoleon' in Schulze and Etienne François (eds), *Deutsche Erinnerungsorte*, vol. 2 (Munich: 2002), pp. 28–46; Friedrich Stählin, *Napoleons Glanz und Fall im deutschen Urteil: Wandlungen des deutschen Napoleonbildes* (Braunschweig: 1952).
15. See Christian Koller, *Fremdherrschaft: Ein politischer Kampfbegriff im Zeitalter des Nationalismus* (Frankfurt/M–New York: 2005), pp. 198–381; Koller, '"Die Fremdherrschaft ist immer ein politisches Uebel" – Die Genese des Fremdherrschaftskonzepts in der politischen Sprache Deutschlands im Zeichen umstrittener Herrschaftslegitimation' in Helga Schnabel-Schüle and Andreas Gestrich (eds), *Fremde Herrscher – fremdes Volk: Inklusions- und Exklusionsfiguren bei Herrschaftswechseln in Europa* (Frankfurt: 2006), pp. 21–40.
16. See Johann Ludwig Klüber, *Wichtige Urkunden für den Rechtszustand der deutschen Nation: Mit eigenhändigen Anmerkungen: Aus seinen Papieren mitgeteilt und erläutert von Karl Theodor Welcker*, 2nd edn (Mannheim: 1845), p. 3; *Die Revolution und die Revolutionäre in Italien* (Leipzig: 1846), p. 5; *Allgemeine Zeitung*, 29 April 1848, 23 November 1848; 'Die Oesterreicher in Italien und die italienische Politik Russlands III: Von 1815 bis auf die Gegenwart', *Preussische Jahrbücher* 2 (1858), pp. 268–303, here: 274; 'Die italienische Frage', *Grenzboten* 18/1/2 (1859), pp. 61–70/175–85, here: 63, 65, 67; H.[einrich] B.[ernhard] Oppenheim, *Deutsche Begeisterung und Habsburgischer Kronbesitz* (Berlin: 1859), p. 26; [Constantin Rössler], *Preussen und die italienische Frage* (Berlin: 1859), pp. 15, 16, 18, 20.
17. See 'Beiträge zur Geschichte Irlands: Siebenter Artikel', *Historisch-politische Blätter für das katholische Deutschland* 12 (1843), 618–31, here: 618.
18. See Theodor Mommsen, *Römische Geschichte*, vol. 3 (Leipzig: 1856), p. 759; Wilhelm von Giesebrecht, *Geschichte der deutschen Kaiserzeit*, vol. 1, 4th edn (Braunschweig: 1873), p. 20; Felix Dahn, *Urgeschichte der germanischen und*

romanischen Völker, vol. 2 (Berlin: 1881), p. 126; Dahn, *Armin der Cherusker: Erinnerungen an die Varus-Schlacht im Jahre 9 nach Chr* (Munich: 1909), pp. 9, 17.
19. See Johann Gustav Droysen, *Geschichte der Preussischen Politik*, part 2, vol. 3/1 (Leipzig: 1863), p. 4.
20. On this double function of political concepts as indicators as well as factors of changes in political consciousness, see Reinhart Koselleck, 'Einleitung' in Reinhart Koselleck, Otto Brunner and Werner Conze (eds), *Geschichtliche Grundbegriffe: Historisches Lexikon zur politisch-sozialen Sprache in Deutschland*, vol. 1 (Stuttgart: 1972), pp. XIII–XXVII; Koselleck Reinhart, *The Practice of Conceptual History: Timing History, Spacing Concepts* (Stanford: 2002); Rolf Reichardt, 'Einleitung' in Rolf Reichardt and Eberhard Schmitt (eds), *Handbuch politisch-sozialer Grundbegriffe in Frankreich 1680–1820*, vol. 1/2 (Munich: 1985), pp. 39–148.
21. See Johann Gustav Droysen, *Vorlesungen über das Zeitalter der Freiheitskriege*, vol. 1, 2nd edn (Gotha: 1886), pp. 10–11; Theodor Rohmer, *Deutschlands Beruf in der Gegenwart und Zukunft* (Zurich-Winterthur: 1841), p. 51; J.[ohann] G.[eorg] A.[ugust] Wirth, *Die Rechte des deutschen Volkes*. s. l (1838), p. 106; *Deutsche Zeitung*, 19 March 1848.
22. See Johann Gustav Droysen, *Vorlesungen über das Zeitalter der Freiheitskriege*, vol. 1, 2nd edn (Gotha: 1886), pp. 10–11.
23. Franz Otto and Ed Grosse (eds), Vaterländisches Ehrenbuch: Grosse Tage aus Preussens und Deutschlands Geschichte: Gedenkbuch an die glorreichen Jahre 1813 bis 1815, 3rd edn (Berlin-Leipzig: 1870), p. 38.
24. Roderich Benedix, *1813. 1814. 1815: Volksbuch* (Wesel: 1842), pp. 479–80.
25. Theodor Rohmer, *Deutschlands Beruf in der Gegenwart und Zukunft* (Zurich-Winterthur: 1841).
26. Fr.[iedrich] Ahlfeld, *Danket dem Herrn, dem grossen Siegverleiher: Predigt über Psalm 46, 8–12 am 50jährigen Jubiläum der Leipziger Völkerschlacht den 18. October 1863 in der Kirche zu St. Nicolai in Leipzig* (Leipzig-Dresden: 1863), p. 4.
27. B. B. Brückner, *Die Befreiung des deutschen Vaterlandes: Predigt bei der Gedenkfeier der Leipziger Völkerschlacht am 20. Sonntag nach Trinitatis 1863* (Leipzig: 1863), p. 320.
28. *Deutsche Allgemeine Zeitung*, 11 May 1848.
29. *Allgemeine Zeitung*, 25 October 1863.
30. Carl Follenius, 'Ueber die Gründe der Untersuchung demagogischer Umtriebe und Verschwörungen in Deutschland' in Johannes Wit, genannt von Düring, *Fragmente aus meinem Leben und meiner Zeit*, vol. 3/1 (Leipzig: 1828). pp. 177–93, here: 182.
31. Ludwig Börne, *Gesammelte Schriften*, vol. 11 (Hamburg-Frankfurt/M: 1862), p. 34.
32. *Neue Rheinische Zeitung*, 30 July 1848.
33. Karl Marx and Friedrich Engels, 'Die deutsche Ideologie: Kritik der neuesten deutschen Philosophie in ihren Repräsentanten Feuerbach, B. Bauer und Stirner, und des deutschen Sozialismus in seinen verschiedenen Propheten' [1846] in *Marx-Engels-Werke*, vol. 3, pp. 9–530, here: 179; [Friedrich Engels], 'Deutscher Sozialismus in Versen und Prosa' [1847] in *Marx-Engels-Werke*, vol. 4, pp. 207–47, here: 233.
34. See, for instance, Theodor Lindner, *1813* (Halle/S: 1914), p. 32.

35. See, for instance, *Stenographische Berichte über die Verhandlungen des Reichstags*, XI. Legislaturperiode, II. Session 1905/1906, vol. 1 (Berlin: 1906), p. 241.
36. See, for instance, *Germania*, 8 March 1913.
37. See, for instance, K.[arl] Th.[eodor] Heigel, *Deutsche Geschichte vom Tode Friedrichs d. Gr. bis zur Auflösung des alten Reichs*, vol. 1 (Stuttgart: 1899), p. v; Gerhard Ritter, *Die preussischen Konservativen und Bismarcks deutsche Politik 1858–1876* (Heidelberg: 1913), p. 2; *Kölnische Zeitung*, 18 October 1913.
38. See, for instance, Richard Schwemer, *Restauration und Revolution: Skizzen zur Entwicklungsgeschichte der deutschen Einheit* (Leipzig: 1902), pp. 15–16.
39. Friedrich Meinecke, 'Fichte als nationaler Prophet' (1908) in Meinecke (ed.), *Preussen und Deutschland im 19. und 20. Jahrhundert: Historische und politische Aufsätze* (Munich-Berlin: 1918), pp. 134–49, here: 134.
40. Heinrich Ulmann, *Geschichte der Befreiungs-Kriege 1813 u. 1814*, vol. 1 (Berlin: 1914), pp. 42–3.
41. *Kölnische Zeitung*, 11 March 1913.
42. *Germania*, 18 October 1913.
43. See, for instance, Jörn Rüsen, *Konfigurationen des Historismus: Studien zur deutschen Wissenschaftskultur* (Frankfurt/M: 1993); Wolfgang Hardtwig, 'Von Preussens Aufgabe in Deutschland zu Deutschlands Aufgabe in der Welt' in Hardtwig (ed.) *Geschichtskultur und Wissenschaft* (Munich: 1990), pp. 103–60.
44. See, for instance, Heinrich von Treitschke, 'Der Krieg und die Bundesreform' [25. Mai 1866], in Treitschke (ed.), *Zehn Jahre Deutscher Kämpfe 1865–1874: Schriften zur Tagespolitik* (Berlin: 1874), pp. 67–99; Treitschke, 'Die Zukunft der norddeutschen Mittelstaaten' [30. Juni 1866], in Treitschke (ed.), *Zehn Jahre Deutscher Kämpfe 1865–1874: Schriften zur Tagespolitik* (Berlin: 1874), pp. 8–26; Treitschke, *Deutsche Geschichte im Neunzehnten Jahrhundert*, vol. 1 (Leipzig: 1879); Heinrich von Sybel, *Die Deutsche Nation und das Kaiserreich: Eine historisch-politische Abhandlung* (Duesseldorf: 1862).
45. August Bebel, *Die Frau und der Sozialismus: Mit einem einleitenden Vorwort von Eduard Bernstein* [1929] (Bonn: 1994), p. 112.
46. Franz Mehring, 'Jena und Tilsit: Ein Kapitel ostelbischer Junkergeschichte' (1906), in Mehring and Gesammelte Schriften (eds) *Gesammelte Schriften*, vol. 6 (Berlin: 1976), pp. 7–151, here: 147–8. Similarly: *Vorwärts*, 14 October 1906.
47. Franz Mehring, 'Jena (3. Oktober 1906)' in Mehring and Gesammelte Schriften (eds) *Gesammelte Schriften*, vol. 6 (Berlin: 1976), pp. 160–3, here: 161.
48. Kurt Eisner, *Das Ende des Reichs: Deutschland und Preussen im Zeitalter der grossen Revolution* (Berlin: 1907), pp. 346–7.
49. *Vorwärts*, 27 February 1913.
50. See *Vorwärts*, 17, 19 & 30 October 1913.
51. See, for instance, Dieter Düding, 'Antisemitismus als Parteidoktrin: Die ersten antisemitischen Parteien in Deutschland (1879–1894)' in Harm Klueting (ed.), *Nation, Nationalismus, Postnation: Beiträge zur Identitätsfindung der Deutschen im 19. und 20. Jahrhundert* (Cologne: 1992), pp. 59–70; Peter G. Pulzer, *The Rise of Political Anti-Semitism in Germany & Austria* (New York-London-Sydney: 1964); Richard S. Levy, *The Downfall of the Anti-Semitic Political Parties in Imperial Germany* (New Haven-London: 1975); Kurt Wawrzinek, *Die Entstehung der deutschen Antisemitenparteien (1873–1890)* (Berlin: 1927).

52. See Moshe Zimmermann, *Wilhelm Marr: The Patriarch of Anti-Semitism* (New York-Oxford: 1986).
53. W.[ilhelm] Marr, *Der Sieg des Judenthums über das Germanenthum: Vom nicht confessionellen Standpunkt aus betrachtet* (Berne: 1879), p. 32.
54. *Allgemeine Evangelisch-Lutherische Kirchenzeitung*, 10, 24 & 31 December 1880; 14 January 1881.
55. *Stenographische Berichte über die Verhandlungen des Reichstags*: VIII. Legislaturperiode, I. Session 1890/92, vol. 5, p. 3586.
56. Shulamit Volkov, 'Antisemitismus als kultureller Code' in Volkov (ed.), *Antisemitismus als kultureller Code: Zehn Essays*, 2nd edn (Munich: 2000), pp. 13–36.
57. See also Christian Koller, 'Fremdherrschaft und nationale Loyalität: Das Fremdherrschaftskonzept in der politischen Sprache Deutschlands der ersten Hälfte des 20. Jahrhunderts' in Joachim Tauber (ed.), *Kollaboration in Nordosteuropa: Erscheinungsformen und Deutungen im 20. Jahrhundert* (Wiesbaden: 2006), pp. 56–74; Heinz Sproll, *Französische Revolution und Napoleonische Zeit in der historisch-politischen Kultur der Weimarer Republik: Geschichtswissenschaft und Geschichtsunterricht 1918–1933* (Munich: 1992).
58. See, for instance, *Tage-Blatt für den Kanton Schaffhausen* (20 July 1922) regarding a speech of General von Einem.
59. Max von Szczepanski, 'Die Lehre von Leipzig', *Grenzboten* 78/4 (1919), 49–53, here: 53.
60. Gerhard L. Weinberg (ed.), *Hitlers zweites Buch: Ein Dokument aus dem Jahr 1928* (Stuttgart: 1961), pp. 146–7.
61. *Neue Preussische Zeitung*, 30 March 1923.
62. See Armin Mohler, *Die konservative Revolution in Deutschland 1918–1932: Ein Handbuch*, 2 vols 3rd edn (Darmstadt: 1989); Stefan Breuer, *Anatomie der konservativen Revolution* (Darmstadt: 1993); Heide Gerstenberger, *Der revolutionäre Konservatismus: Ein Beitrag zur Analyse des Liberalismus* (Berlin: 1969).
63. [Arthur] Moeller van den Bruck, *Das dritte Reich*, in Hans Schwarz (ed.), 3rd edn (Hamburg: 1931), pp. 184–5.
64. See also Christian Koller, 'French Revolution' in Cyprian P. Blamires (ed.), *World Fascism: A Historical Encyclopedia* (Santa Barbara: 2006), pp. 257–9.
65. *Germania*, 31 March 1935.
66. Hans Erich Feine, *Das Werden des Deutschen Staates seit dem Ausgang des Heiligen Römischen Reiches, 1800 bis 1933: Eine verfassungsgeschichtliche Darstellung* (Stuttgart: 1936), p. 48.
67. See, for instance, Eberhard Jäckel and Alex Kuhn (eds), *Hitler: Sämtliche Aufzeichnungen 1905–1924* (Stuttgart: 1980), pp. 169–70 and 176; Adolf Hitler, *Mein Kampf: Zwei Bände in einem Band* 676–680th edn (Munich: 1941), p. 711; Gerhard L. Weinberg (ed.), *Hitlers zweites Buch: Ein Dokument aus dem Jahr 1928* (Stuttgart: 1961), p. 193; Karl Siegmar Baron von Galéra, *Deutsche unter Fremdherrschaft: Die Geschichte der geraubten und unerlösten deutschen Gebiete* vol. 2 (Leipzig: 1933), pp. 420–1.
68. See, for instance, Eberhard Jäckel and Alex Kuhn (eds), *Hitler: Sämtliche Aufzeichnungen 1905–1924* (Stuttgart: 1980), p. 267; Alfred Rosenberg, 'Jüdische Zeitfragen' [1919] in Rosenberg, *Blut und Ehre: Ein Kampf für deutsche Wiedergeburt: Reden und Aufsätze von 1919–1933*, ed Thilo von Trotha (Munich: 1934), pp. 15–27, here: 27; Rosenberg, *Die Protokolle der Weisen von*

Zion und die jüdische Weltpolitik 2nd edn (Munich: 1923), S. 141; Rosenberg, *Wesen, Grundsätze und Ziele der Nationalsozialistischen Deutschen Arbeiterpartei: Das Programm der Bewegung* (Munich: 1933), p. 24.

69. See, for instance, *Völkischer Beobachter*, 23 December 1934, 13 November 1935, 4 October 1938, 18 October 1938; Heinrich Schneider (ed.), *Unsere Saar* (Berlin: 1934), pp. 3, 17; *Germania*, 6 January 1935, 9 January 1935, 1 March 1935; *Kölnische Zeitung*, 7 January 1935, 10 January 1935; *Frankfurter Zeitung*, 8 January 1935; Friedrich Grimm, *Frankreich an der Saar: Der Kampf um die Saar im Lichte der historischen französischen Rheinpolitik* (Hamburg: 1934), p. 38; Karl Siegmar Baron von Galéra, *Österreichs Rückkehr ins Deutsche Reich: Von Kaiser Karl zu Adolf Hitler* (Leipzig: 1938), p. 91; Günter Wüster, 'Uns ruft die Ostmark' in *Des Führers Wehrmacht half Grossdeutschland schaffen: Berichte deutscher Soldaten von der Befreiung der Ostmark und des Sudetenlandes* (Berlin: 1939), pp. 27–34, here: 33.
70. G. Hoffmann, *Von deutscher Freiheit: 5 Jahrhunderte deutscher Geschichte* (Prague: 1937).
71. Hermann Rauschning, *Die Revolution des Nihilismus: Kulisse und Wirklichkeit im Dritten Reich* 3rd edn (Zurich-New York: 1938), p. 157.
72. See An die Jugend Oesterreichs! [1938]; 'Oesterreich: "Für Oesterreichs Freiheit und Unabhängigkeit!"' in *Rundschau* 7 (1938), p. 1313; *Oesterreich unter dem Reichskommissar: Bilanz eines Jahres Fremdherrschaft* (Paris: 1939); 'Der Kampf um die Befreiung Oesterreichs von der Fremdherrschaft: Resolution des Zentralkomitees der Kommunistischen Partei Oesterreichs' in *Rundschau* 8 (1939), pp. 1473–6; Johann Koplenig, *Trotz alledem: Oesterreichs Volk kämpft weiter für seine Unabhängigkeit* in *Rundschau* 7 (1938), pp. 547–8; Fritz Alt, *Schmücke dein Heim!* (Leipzig: s.d.); Heinrich Baumann, 'Österreich unter dem Reichskommissar' *in Weg und Ziel* 3 (1938), pp. 220–4; 'Für die Einheitspartei der österreichischen Arbeiterklasse!' in *Weg und Ziel* 3 (1938), pp. 210–15, here: 214; Peter Wieden, *Arbeiterklasse und Nation* [1939].
73. *Völkischer Beobachter*, 3 April 1945.
74. Friedrich Meinecke, *Die deutsche Katastrophe: Betrachtungen und Erinnerungen* (Wiesbaden: 1946), pp. 151–2.

4

From Heroic Defeat to Mutilated Victory: The Myth of Caporetto in Fascist Italy

Vanda Wilcox

Defeat has been synonymous with the Italian army in European popular perceptions for much of the twentieth century, but stereotypes of Italian military ineptitude abounded within Italy before unification. Indeed many contemporary supporters of the Risorgimento were keen to procure unification by force of arms rather than through negotiation explicitly to prove that Italians could and would fight; it was feared that Italians were innately unsoldierly.[1] Mazzini, for instance, was concerned that the Italian 'vices' of 'sentimentality and morbid compassion' would hinder the nation from obtaining its true greatness.[2] Others agreed: in 1870 the French theorist Ardant du Picq, who had served in Italy, wrote that 'Italy will never have a really firm army. The Italians are too civilized, too fine'.[3] The defeats and humiliations of the half-century preceding the First World War, especially against African opponents, had created considerable fears among some Italians that as a people they lacked martial qualities. Supporters of intervention in the First World War were keen to display Italy's martial abilities not only to themselves but to the world, and to expunge the humiliations of her defeats against Austria-Hungary at Custoza and Lissa (1866) and especially the colonial disasters of Dogali (1887) and Adowa (1896).[4] Catastrophic defeat, and the consequent fear of future defeat, was thus a well-established trope in Italian culture by 1915. In a sense, the seeds of Caporetto – and the way it was to be remembered – were sown in the national psyche long before the war itself began.

The battle of Caporetto occupied a curious position in post-war Italian memory and debate. It was something of a paradox, a terrible defeat which had led to ultimate victory – but that victory itself was equally problematic, since Italy's 'betrayal' at Versailles was a kind of defeat. In the years following the war, a rhetorical inversion in public discourse led to the presentation of the defeat of Caporetto as a kind of victory, and of the victory of the war as a defeat.

The memory of the First World War in Italy ever since the Armistice has been highly diverse and varied. There was no single 'war experience', even for mobilized men, but rather a complex and highly diverse range of experiences dependent on social, regional and political differences and on the variety of different types of wartime service. For civilians, war experiences were further shaped by differences of gender, urban or rural life and age. The memory of the war must be considered as a textured picture of varying interpretations, both emotional and political in origin. This chapter will focus on developments in the formation of memory in the immediate post-war period and under the Fascist regime that followed. It will examine the construction of memory through political debate, literary and historical writings, monuments and memorials, and battlefield pilgrimage. War literature was highly influential in the development of collective memory in post-war Italy (as elsewhere) and also frequently shaped the ways in which individuals recollected their own experiences in the years to come.[5] Monuments and memorials represented a concrete expression of the interpretation of the war by those who erected them – chiefly, in this case, the Fascist regime and its local agents.[6] Collectively these sources permit an examination of the ways in which Caporetto was remembered, interpreted and understood in the years before World War II.

From defeat to victory

The battle of Caporetto began in the early hours of 24 October 1917, with a joint German–Austrian attack at the top of the Isonzo front, around the small town of Caporetto. Brief but thorough bombardments followed by the effective use of gas enabled German and Austrian troops to break through the Italian lines at Tolmino and at Conca di Plezzo. The Italian troops, already exhausted, demoralized and poorly equipped, were taken by surprise and rapidly defeated by the unfamiliar infiltration tactics of the elite German *Jäger* and *Sturm* battalions. The vigour of the resistance varied: while some units fought bravely until their defeat was complete and unavoidable, others surrendered almost immediately or fled in panic. Though excellent intelligence on the location, timing and objectives of the attack had been available to Italian senior commanders, they had chosen to disregard these reports almost entirely. Poor organization and inadequate planning meant that there was an almost total absence of reserves in this sector, and those that were available were not appropriately deployed.

In the course of the initial battle, the entire left flank of General Luigi Capello's II Army was forced to abandon its positions, and soon the whole army was in retreat. Within a few days III Army, stationed on the Isonzo to the south, was also forced to retire. Supreme Command abandoned its headquarters at Udine on 27 October and ordered a full retreat to the line of the Tagliamento River. On 30 October the retreating troops began to blow up the bridges across the Tagliamento, abandoning not only fleeing civilians but entire Italian army corps on the other side of the river. As the enemy continued to advance, succeeding in crossing the river in several places, it became apparent that this new line would not suffice, and a withdrawal to the Piave was announced on 4 November. IV Army was also ordered to withdraw from the Cadore sector and re-establish itself on the river Piave. By the end of November, when the Italian front line stabilized, the army had retreated over 150km, abandoning 20,000 square kilometres of territory to the Austrians. II Army had totally disintegrated and its 670,000 men scattered: over 280,000 of them had been taken prisoner and a further 350,000 had absconded, while 40,000 were killed or wounded. In addition, 400,000 civilians had fled the newly occupied region. Artillery and essential equipment were also lost – over 3000 artillery pieces and a similar number of machine guns were abandoned to the advancing Austrians along with huge quantities of munitions, food, animal fodder, uniforms and vital medicines.[7]

Caporetto has long been perceived both within Italy and abroad as the key moment of the war, as the vital turning point in a variety of respects.[8] Symbolic discontinuity is at the heart of this view, as the abandonment of the old front along the Carso and the establishment of a new physical and conceptual theatre of war – the Piave – emphasized the change which had taken place. The reforms carried out under the new commander in chief, General Armando Diaz, made the army more effective. Tactics and training were improved as were administration, leadership and man-management, with a substantial impact on both morale and on battlefield effectiveness.[9] Thus improved, the Italian army ably defeated the Austrian offensive at the battle of the Piave in June 1918, and then proceeded to its final successful offensive at Vittorio Veneto in 1918. By the Armistice the Italians had recrossed the Tagliamento, but were still some way from regaining all the ground lost during their great defeat.

Caporetto thus heralded the beginning of a programme of military reform. Arguably though its greatest effect was political: it laid bare the tensions inherent between state and society, people and government and the flaws which arose from them in both civic and military governance. Certainly the predominant modes of contemporary interpretation of the defeat rejected military explanations in favour of the political.

For the Right, Caporetto was an indictment of the liberal state, a clear failure of the democratic regime which thus legitimized the fascist revolution. Mass defeatism was the direct consequence of Giolittian neutralism. For many on the Left, Caporetto was a failed revolution – a spontaneous mass rising, which had unfortunately failed to achieve a change in regime.[10] Accounts of mass surrender and desertion were thus presented not as cowardice or disorder but the concerted and deliberate actions of the enlightened masses. The radical left-wing polemicist (and future Fascist) Curzio Malaparte described it as 'the bravest, most wonderful action of the proletariat under arms'.[11] More moderate leftists viewed Caporetto as a military strike – less threatening than a revolution, which evoked Bolshevik overtones, but nonetheless an organized movement of the working classes against oppression. All these interpretations were primarily political: Caporetto rendered tangible the conflict between 'legal' and 'real' Italy.[12] More recent historiographical explanations of the defeat have chiefly focused on military factors. Caporetto represented both, in the words of Bencivenga, a 'strategic surprise' and a set of tactical innovations. The social and political circumstances served both to facilitate the Austrian advance and to render the defeat more serious, but did not, in this view, actually cause the crisis.[13] The interpretative battle, though, was only a precursor to the construction of memory. Contemporary political explanations could not undo the catastrophic nature of the battle nor could they provide a fully reassuring national myth, and this was largely to do with the crushing disappointments the nation was to suffer in 1919. The heroism of the defeat had mutated into what was famously described by the poet Gabriele D'Annunzio as a mutilated victory.

The mutilated victory

The potency of the defeat was all the greater given that in the aftermath of the war, it seemed that Italy's victory had been snatched from her. In negotiations at Versailles Italy was not rewarded with all the land she had been promised, nor did she benefit economically as other nations were perceived to have done.[14] It was widely felt in Italy that the severity of Italian sacrifices were wholly disproportionate with the tiny gains the nation had made in return. Her efforts for the Entente, it appeared, were not appreciated – which caused great resentment of these former allies.[15] Not only did Italy not receive as much territory as she had hoped but many veterans could not see in what way the new land was an advantage; it seemed that despite victory the war had only damaged the nation. In the reminiscence of one peasant veteran,

> The war of '18 left only wretchedness, they conquered a bunch of mountains and a pile of bloody rocks . . . well, obviously I know nothing about politics.[16]

The expedition to Fiume was one result of the 'mutilated victory', as unsatisfied irredentists took the law into their own hands in the effort to obtain what they believed to be the nation's just reward. Since the treaty of Versailles had provided no recompense for Caporetto, yet another victory was required. Vittorio Veneto, it seemed, was not sufficiently important to have wiped out the 'stain' of Caporetto.[17] Another result of having 'won the war but lost the peace' was the urgent need to find an acceptable national interpretative framework for the war – and especially for the defeat at Caporetto. If Italy was to be successfully presented as an unfairly victimized victorious state, her moment of crisis had to be explained away. Of course, pre-existing Italian anxieties over military performance, and the shame of past humiliations, invested the national post-mortem on Caporetto with added significance.

As in many crises, the almost universal initial response was to find scapegoats and allocate responsibility for the disaster.[18] The general staff struck the first blow in this interpretative battle. On 28 October, Cadorna issued his infamous bulletin in which he denounced the troops of II army, blaming their 'lack of resistance' (*mancata resistenza*) for the terrible defeat. Although the government did not publish the text in full, word soon spread of the commander-in-chief's allegations, which served only to alienate even further the demoralized troops. As one peasant soldier commented, 'after Caporetto [the army] called us every sort of savage, saying "cowards, traitors!"'.[19]

This attempt to blame the rank and file for the defeat was politically disastrous for the commander-in-chief and made his dismissal inevitable;[20] it seemed that the only alternative, though, was to hold Cadorna and Supreme Command responsible.[21] This, broadly, was the conclusion of the Royal Commission of Inquiry into Caporetto whose report was published in August 1919. The report was highly critical of Cadorna, Capello and other generals, while rejecting completely allegations of an antimilitarist plot.[22] The work of the commission was flawed in several ways. Crucially it was hampered by the appointment of one of the chief targets of Cadorna's criticism, Orlando, as the new Prime Minister. This meant that the former chief of staff's allegations of government inefficiency and defeatism were not properly investigated.[23] Since blame could not be attached to the government by this government-run investigation, there remained only the generals who could safely be held responsible without further alienating the population. Junior officers were also inclined to attribute much of the blame to Supreme Command; to senior officers at army, corps and divisional level; and to the general staff.[24] Valentino Coda blamed the chaotic and disorganized nature of the defeat on poor leadership and command, and was particularly critical of Cadorna himself, while Luigi Gasparotto was highly critical of the general staff and their planning.[25]

But the simplistic dichotomy which emerged in 1918–19 – blaming either the mismanagement and ineptitude of Supreme Command and the generals or the betrayal and sedition of the troops – could not long be sustained, especially after 1922. Fascist rhetoric required both that Supreme Command and the generals retained a high reputation, and that the ordinary soldiers' experiences were celebrated.[26] Thus the dichotomy of analysis most common in the immediate post-Caporetto period could no longer be used: but how, then, was it to be explained or remembered?

The simplest option, politically, was not to remember the defeat at all. By the time the report of the inquiry commission was published, Prime Minister Nitti commented that Vittorio Veneto had wiped the slate clean; he announced that 'victory has healed everything . . . he who wins is right'.[27] These efforts to suppress debate were surprisingly successful, in that Caporetto ceased to be a major political issue by 1920, with even the Italian Socialist Party abandoning discussion of the defeat. Once Caporetto no longer belonged in the political arena, it was open instead to the process of mythologization.[28] For Mussolini it became axiomatic that Caporetto had been nothing but a parenthesis, and was now a closed and no longer relevant episode.[29] Though on the first anniversary of the battle in October 1918 he had called for further investigation into the defeat, claiming that it must not be forgotten or excluded from political debate, once he came to power himself he abandoned this potentially inconvenient position.[30] He aimed to heal the rifts within army circles and in society as a whole, in pursuit of which objective, in 1924, he elevated both Cadorna and Diaz to the status of Marshal of Italy.

The defeat could be downplayed: 'the retreat of Caporetto did not halt the progress of Italian arms, but was merely a brief episode', explained one guidebook.[31] When Angelo Gatti requested access to cabinet papers and other government documents in order to write his own history of the battle, he was denied permission. Mussolini told him, 'Now is not the time for history. This is the time for myths.'[32] Mussolini's aim was not to remember but to forget, and he was not alone: many veterans sought to forget the war and leave behind the mental trauma they had suffered.

But it was not possible at a national level simply to forget a battle of this magnitude – however desirable it might seem. Too many individuals had been involved, and the event was too significant. Therefore it must be made 'acceptable' and a new, patriotic narrative of Caporetto be established. One solution was to blame the liberal government and civilian shirkers for undermining patriotic loyalties and national determination.[33] This was a pre-existing trope much used in 1914–15 during the debate on intervention, when Giolitti and the various neutralist interest groups were labelled traitors. Gioacchino Volpe blamed socialists and liberals for the defeat, claiming that they had laid the foundations for the crisis.[34] Malaparte also blamed Italian civilians, whom he compared to the more committed populations of France and Germany, for undermining the soldiers' sacrifices and betraying the army's endeavours.[35] But unlike in Germany, where the stab-in-the-back myth relied upon the idea of being 'undefeated' on the battlefield, this model could not explain the defeat's military manifestation.

Individuals and small- or medium-sized units could comfortably construct narratives of the battle which placed the blame elsewhere upon unspecified nameless others. Rather than omit or minimize the episode, the official regimental history of the thirtieth field artillery included a detailed description of the defeat – yet without apportioning any blame or criticizing anyone. It suggests that they at least had been nothing but heroic and determined:

> Our officers, even on this humiliating if memorable day, wrote a golden page in the history of our brave regiment.[36]

Peasant veterans in recounting their experiences were also keen to emphasize individual valour:

> Our division resisted for 2 days, [but] our artillery was already firing towards our rear lines, behind us, so we saved ourselves as best we could, we fled.[37]

Here it is the artillery who are blamed for rendering the infantry's position untenable. Predictably, self-exoneration is a recurrent feature of the war literature: veterans of III Army described themselves as 'invitti' (undefeated), emphasizing that their withdrawal had been forced upon them by the retreat of II Army, not by their own defeat.[38] Even Malaparte, who was so proud to declare himself a 'caporettist', was at pains to point out that he himself was stationed in the high Cadore with IV Army and was thus not himself personally involved in the defeat.[39] Officers of II Army also found ways to distance themselves from the disaster. Ardegno Soffici, who was attached to II Army during the retreat, emphasized that he participated merely as an observer and message bearer, and had no personal responsibility.[40] Valentino Coda stressed that he was not present at Caporetto itself, but in quite another sector of the line.[41]

But for a truly national myth, the tactic of simply shifting the blame onto some nameless other could not succeed. Since neither military nor political scapegoats were available which would fully explain the defeat, the emphasis had to be diverted from blame towards reintegration of the event. An inclusive myth was required which could encompass the entire nation in the same way as the totalizing discourse of the war itself. Hence an accepted narrative emerged which did not *deny* the defeat, but recast it as the progenitor of renewal. The story of Caporetto was retold to illuminate the positive outcomes of the moment of national trauma.

Defeat and rebirth

Wolfgang Schivelbusch has shown how the classical model of the fall of Troy, together with the Christian paradigm of sacrifice and resurrection, have provided an enduring interpretative framework for defeat in Western societies; as he writes, 'defeated nations have a perennial identification with the Passion of Christ.'[42] Italy in 1918–9 was not a defeated nation; but she was a nation who had suffered a great defeat, and whose subsequent victory had failed to produce the expected benefits. The triumphalism of victory, in the memory of the war, was thus combined with the anxieties of defeat. Since the 'mutilated victory' was no great cause for celebration, and did not bear too much close scrutiny, other foci had to be sought.

Thus the *response* to the defeat was something that could be celebrated: under which circumstances, there was no need to downplay the scale of the initial disaster, since the greater the suffering of the nation, the greater the army's achievement in overcoming its trauma. In this officially sanctioned narrative, the successful conduct of the battle of the Piave in June 1918 'tore up and destroyed the sad page of Caporetto', leaving Italy 'safe, and ready for new trials'.[43] This argument, provocatively supported by historian Lucio Ceva who claimed that Caporetto 'became an extremely costly Italian success',[44] can be applied to both military and political developments. The shorter front made for more efficient communications and easier deployment of reserves, while Diaz's generalship was of a higher quality in strategic and operational respects. Socially, the trauma of occupation by Austria, the old enemy, and the vigorous resistance on the Piave led to an unprecedented outpouring of patriotic fervour and national determination which was crucial in the pursuit of victory. Politically, a truce between various factions was established for the first time, albeit temporarily, in the face of the catastrophe. Orlando's new government minimized dissent and achieved an unprecedented degree of centralized control and cooperation.[45] In the words of Giuseppe Prezzolini, 'Italy was united as it had not been for centuries – in fact, in law and in common consciousness.'[46]

Caporetto, then, could be presented as a moment of national renewal. Indeed it was repeatedly claimed by the Fascist party as the moment of their birth, and all aspects of the memory and commemoration of the war underwent a process of fascistization.[47] Malaparte wrote, 'Caporetto was the first sign of a new direction for humanity. All rebirths [renaissances] are Italian.'[48] In the introduction to his 1923 edition of *Viva Caporetto* – designed to conform more closely to the ideals of the new regime – he emphasized that the 'revolution' of Caporetto has reached its triumphant conclusion with the March on Rome. As Schivelbusch observes, the idea of defeat as the birth of a new political system which embodies the will of the people and renews their patriotic fervour is a common one; in particular, defeat is easily portrayed as a purging process, eliminating undesirable or 'weak' elements to make way for the new.[49]

Caporetto could also be interpreted as a trial of faith. Much was made of Mussolini's 'unwavering faith' in Italian victory[50] throughout the retreat, while at the time, doubts about Italian ability in the aftermath of the defeat were labelled treasonous. The maintenance of obedience and discipline among the troops were not just a source of pride in themselves but a sign of continued belief in victory.[51] Mario Puccini, writing about Caporetto before the end of the war, concluded with a bloodthirsty determination to eradicate Austrian occupation and his certainty of victory: we will return to the Carso, he wrote, drawing confidence from the orderly and disciplined behaviour of III army during the retreat.[52] Caporetto was thus a successful triumph over despair.

Memorials: Physical manifestations of memory and commemoration

In Italy the memory of the war was highly topographically orientated: symbolic locations formed a key part of the discussion and conceptualization of the war in the post-war years. Veterans had commonly served in many parts of the front and had personal recollections of the now-iconic battlefields. But those who had not served could also develop a link to these semiotically charged locations. Battlefield pilgrimage was a booming industry in the 1920s, much encouraged by the Fascist party, and described in one guidebook as a 'pious and sacred . . . duty for every Italian'.[53] Trips were organized by the party-run Dopolavoro organizations for adults and through schools and universities for the young. From 1927 the *Touring Club Italiano* published a seven volume series of battlefield guides which ran to five or six editions each.[54] Trips to visit cemeteries, monuments and preserved trenches were extremely popular: the cemetery of Redipuglia, perhaps the most important site, was receiving over 100,000 visitors annually by the mid-1920s.[55]

Visitors also travelled to the newly Italian town of Caporetto where in 1938 a huge monument was erected at the behest of the central government, and in cooperation with local authorities. The building was both a memorial and a burial ground: the remains of around 7000 Italian soldiers buried in various cemeteries in the locality were disinterred and moved to the Ossuary.[56] This was part of a widespread programme in the 1930s, in which similar Ossuaries were built in dozens of locations, and the bodies of the dead reinterred. Though partly carried out for practical reasons (initial burials were often insufficiently deep to be hygienic), one of the chief aims of this state-directed formal commemoration was the creation of a more homogeneous, Fascist-sanctioned national memory rather than retaining individual local cemeteries with diverse presentations of the war. In these monuments, Fascist iconography is combined with Christian forms of commemoration – part both of the sacralization of public political discourse and of the integration of Catholicism into the new state.[57] Perhaps surprisingly, there is a considerable degree of iconographic consistency between pre-fascist memorials and monuments and those constructed under the regime, though they grew stylistically more homogeneous across the nation from the mid-1920s.[58] The largest shift was in the style and wording of epigraphs on monuments, away from the mourning and grief which characterized immediate responses and towards a celebrative, triumphalist rhetoric which glorified sacrifice in the cause of the nation.[59]

In Caporetto, an already highly significant location for the memory of the war, the rhetoric of heroic defeat is more elaborate than anywhere else. The Ossuary is a fortress-like building on a small hill some 400m above the town, of concentric octagonal terraces ascending towards a small chapel at its centre. This chapel was first consecrated in 1696 to St Anthony, patron of the oppressed and of victims of injustice. The new Fascist memorial was constructed around the old chapel just as Fascist symbolism and iconography were built onto and around the Catholic faith. The *Sacrario* at Caporetto was designed by the architect Giovanni Greppi, with artworks by the sculptor Giannino Castiglioni. This highly celebrated pair received numerous commissions from the Fascist regime, collaborating on the important monuments at Monte Grappa and Redipuglia as well as that at Caporetto.[60]

The monument places prayer at the physical heart of remembrance. It is approached up a steep *via Sacra*, or Sacred Way, built at the same time as the Ossuary, and lined with the 14 Stations of the Cross. The entrance to the Via Crucis is marked by two enormous marble pillars, one embellished with the Cross and the other with the star of Italy, 'the symbols of Patria and Faith',[61] thus creating symbolic unity between Catholicism and militaristic patriotism. Each Station along this Via Crucis is faced by a small piazza where visitors were encouraged to sit and pray or reflect, while contemplating Castiglioni's sculptures. At the top the Ossuary faces onto a 'great piazza where one experiences a vision of mystical solemnity', according to the battlefield guide, where 'the Church watches over its Glorious Martyrs'.[62] The role of the visitor, through prayer and through the experience of climbing the path to the monument, was intended to be active rather than merely passive. For the pilgrim visitor, to mourn the dead of Caporetto was to re-enact Calvary: a symbolic reliving of redemptive sacrifice.

The paradigm of redemptive sacrifice and rebirth is found in other Italian memorials and monuments, yet nowhere else is the connection with the Passion made so clear.[63] The battle of Caporetto is explicitly portrayed in its memorial as the Passion of the nation: the death, resurrection and ultimate triumph of Italy.[64] Likewise for the official Fascist newspaper, *Il Popolo d'Italia*, Caporetto was 'a tremendous Golgotha'.[65] The 'martyrdom' of Cesare Battisti and his fellow irredentists Fabio Filzi and Damiano Chiesa in 1916 formed a part of this same narrative: their death was interpreted as a 'proud confession of the Italian faith'.[66] The Castello di Buon Consiglio in Trento, location of their trial, imprisonment and execution at the hands of the Austro-Hungarians, became a site of mass pilgrimage in its own right as early as July 1919.[67] The redemptive power of both individual and collective sacrifice was demonstrated, in this interpretation, by the vitality and success of the new Fascist Italy – despite the 'betrayal' of Versailles.[68] In enabling this rebirth, the defeat itself is worthy of celebration. In this way the trauma of the mutilated victory was sublimated through the heroism of redemptive defeat.

Wartime sacrifice was not the end, but a beginning: the beginning of a new, better, stronger Italy. Under the provisions of the legislation which established the *Commissario Generale per le Onoranze ai Caduti in Guerra* (general committee to honour those fallen in war), the bodies of the dead remain the responsibility of the state, in perpetuity. They were the 'patrimony of the nation', and they belonged to the state, not their families – who had no say in the location or manner of their burial.[69] In essence, the dead were still in service: above each archway on the Ossuary at Caporetto – as on the massive steps at Redipuglia – is inscribed 'Presente'. This response to the nation or party's summons was a key element of the Fascist liturgy,[70] and in this context a sign of continuing service, even after death. There was no demobilization for the fallen, but rather remobilization in the service of the new national cause – Fascism.

Post-war Italian usages of the word 'Caporetto' offer some suggestions as to the ways in which this diversity of interpretations has endured. Dictionary definitions suggest a shift in the associations of this battle over time. A 1962 definition gives 'caporetto' as a noun or adjective denoting defeat and surrender; but by 1998 this definition had been amended to include the usage guide 'usually humorous'.[71] At 80 years of distance the emotional impact of the battle has diminished to the extent that it can now be treated as a joke or a newspaper headline. The word can be used today to signify some major social disaster such as a long-term economic depression[72] or the collapse and disintegration of an institution or, most commonly, a humiliating defeat in a crucial football match.[73] In particular it can denote a humiliation arising when passive surrender and a lack of fighting spirit have led inexorably to defeat.[74] Pier Paolo Pasolini used the term as a synonym for running away in his 1959 novel about the Roman underclass *Una vita violenta:* 'la maggior parte fece caporetto, squagliandosi per il rione a tutta spinta.'[75] Conversely though, 'the spirit of Caporetto' denotes patriotic fervour and determination,[76]and the possibility of ultimate triumph over adversity.[77] This lasting dichotomy reflects the conflicting narratives of defeat which grew up both officially and informally around the battle, and the enduring inherent contradictions in the social and cultural memory of Caporetto under Fascism. The ennoblement of suffering and the valorization of endurance were combined in a narrative which both reinforced the established cultural norms of Catholicism and bolstered the new regime. By placing both Caporetto and Fascism itself securely into a well-established conceptual framework, the memory of the battle served simultaneously to console, to reassure and to serve as a vital political symbol.

Notes

1. John Whittam, *The Politics of the Italian Army, 1861–1918* (London: 1977), p. 13.
2. Cited in ibid., p. 172.
3. Ardant du Picq, 'Battle Studies' in James Donald Hittle (ed.), *Roots of Strategy* (Mechanicsville: 1987), p. 249.
4. Mario Isnenghi, *Il mito della grande guerra*, 5th edn (Bologna: 2002), pp. 11–19.
5. Mario Isnenghi, *Le Guerre degli italiani: Parole, immagini, ricordi 1848–1945* (Bologna: 2005), pp. 19–20; Nuto Revelli, 'La memoria della guerra nelle campagne cuneesi' in Diego Leoni and Camillo Zadra (eds), *La Grande Guerra: Esperienza, Memoria, Immagine* (Bologna: 1986), p. 609.
6. For general reflections on First World War memorials, see Jay Winter, *Sites of Memory, Sites of Mourning: The Great War in European Cultural History* (Cambridge: 1995), pp. 78–116. For Italian memorials, see Lisa Bregantin, *Caduti nell'obblio: I soldati di Pontelongo nella Grande Guerra* (Portogruaro: 2003); Renato Monteleone and Pino Sarasini, 'I monumenti ai caduti della Grande Guerra' in Leoni and Zadra (eds), *La Grande Guerra*.
7. One of the best sources is the account of Gen. Bencivenga, suppressed by Fascist censors but recently reissued, Roberto Bencivenga, *La sorpresa strategica di Caporetto*, ed. Giorgio Rochat (Udine: 1997). Also of interest is the diary of official historian to the general staff, Angelo Gatti, *Caporetto: Diario di Guerra* (Bologna: 1964), and the account written by Luigi Capello in 1918, Luigi Capello, *Caporetto, perché? La 2. armata e gli avvenimenti dell'ottobre 1917* (Turin: 1967). Excellent recent syntheses can be found in Nicola Labanca, *Caporetto: Storia di una Disfatta* (Florence: 1997) and in the relevant sections of Mario Isnenghi and Giorgio Rochat, *La Grande Guerra, 1914–1918* (Milan: 2000).
8. See Renzo de Felice's introduction in Capello, *Caporetto, perché?*, pp. xi–xiii.
9. On these reforms, see Gian Luigi Gatti, *Dopo Caporetto. Gli Ufficiali P nella Grande guerra: propaganda, assistenza, vigilanza* (Gorizia: 2000).
10. Isnenghi and Rochat, *La Grande Guerra*, p. 384.
11. Curzio Malaparte, *Viva Caporetto! la rivolta dei santi maledetti* (Florence: 1995), p. 51.
12. Isnenghi and Rochat, *La Grande Guerra*, p. 388.
13. See especially Labanca, *Caporetto* and Mario Morselli, *Caporetto: Victory or Defeat?* (London: 2001).
14. For a full analysis of the Conference's implications for Italy and the Italian reaction, see H. James Burgwyn, *The Legend of the Mutilated Victory: Italy, the Great War and the Paris Peace Conference, 1915–1919* (Westport, CT: 1993).
15. Denis Mack Smith, *Italy: A Modern History*, rev. edn (Ann Arbor: 1969), pp. 319–21.
16. Nuto Revelli, *Il mondo dei vinti* (Turin: 1977), p. 73.

17. Mario Isnenghi, *La tragedia necessaria: Da Caporetto all'Otto settembre* (Bologna: 1999), pp. 56–7.
18. Giovanna Procacci, 'The disaster of Caporetto' in John Dickie, John Foot and Frank M. Snowden (eds), *Disastro! Disasters in Italy since 1860: Culture, Politics and Society* (New York: 2002), p. 144.
19. Revelli, *Il mondo dei vinti*, p. 323.
20. Cadorna was dismissed on 8 November by the new prime minister, Orlando, acting not only in line with his own views but also as demanded by Italy's allies. Both Foch and Lloyd George stipulated Cadorna's dismissal as a condition of their sending military support to Italy.
21. On this attitude in the war literature, see Mario Isnenghi, *I vinti di Caporetto nella letteratura di guerra* (Padova: 1967), pp. 56–60.
22. Inchiesta Caporetto, 'Relazione della Commissione d'Inchiesta, 'Dall'Isonzo al Piave 24 ottobre–9 novembre 1917' (Rome: 1919), vol. 2, Conclusion.
23. Isnenghi and Rochat, *La Grande Guerra*, p. 485. Despite exoneration from blame in the commission's report, Orlando sought to delay its publication and minimize public debate and discussion. Labanca, *Caporetto*, pp. 90–1.
24. Carlo Salsa, *Trincee: confidenze di un fante* (Milan: 1995), pp. 314–15; Attilio Frescura, *Diario di un imboscato* (Milan: 1981), pp. 226.
25. Valentino Coda, *Dalla Bainsizza al Piave: all'indomani di Caporetto – appunti d'un ufficiale della II Armata*, 1st edn (Milan: 1919), pp. 62–3, 70–7; Luigi Gasparotto, *Diario di un fante* (Chiari: 2002), p. 175.
26. Pietro Badoglio, Mussolini's chief of the general staff from 1924, was commander of XXVII Corps during the battle of Caporetto, and was at least partially culpable for the defeat in the eyes of many, including the official commission of enquiry. However, the new regime could not, obviously, afford criticism of such a key figure.
27. Cited in Procacci, 'The Disaster of Caporetto', p. 158.
28. Labanca, *Caporetto*, p. 101.
29. Isnenghi and Rochat, *La Grande Guerra*, p. 394.
30. *Il Popolo d'Italia*, 24 October 1918, cited in Isnenghi, *I vinti di Caporetto*, p. 10.
31. Touring Club Italiano (hereafter T.C.I.), *Sui Campi di battaglia del Medio e Basso Isonzo* (Milan: 1927), p. 18.
32. Labanca, *Caporetto*, p. 109.
33. Consociazione Turistica Italiana (hereafter C.T.I.), *Sui Campi di Battaglia: Il Cadore, la Carnia e l'alto Isonzo*, 3rd edn (Milan: 1937), p. 229.
34. Gioacchino Volpe, *Ottobre 1917: dall'Isonzo al Piave* (Rome: 1929), p. 44–5
35. Malaparte, *Viva Caporetto!*, pp. 99–105.
36. 30th Field Artillery Regiment, *Artiglieri del Rubicone: Storia e Memorie* (Rimini: 1937), p. 52.
37. Revelli, *Il mondo dei vinti*, p. 31.
38. *Cimitero Militare di Redipuglia agli Invitti della 3a Armata*, Ediz. Ufficio Centrale Cure Onoranze Salme Caduti in Guerra (Padova: 1920?); see also, Mario Puccini, *Dal Carso al Piave: La ritirata del 3a Armata nelle note d'un combattente*, 1st edn (Florence: 1918).
39. Malaparte, *Viva Caporetto!* p. 192, also introduction by Marino Biondi, pp. 20–21.
40. Ardengo Soffici, *La Ritirata del Friuli: note di un ufficiale della Seconda armata*, 1st edn (Florence: 1919).

41. Coda, *Dalla Bainsizza al Piave*, pp. 6–7.
42. Wolfgang Schivelbusch, *The Culture of Defeat: On National Trauma, Mourning and Recovery* (London: 2003), p. 30.
43. T.C.I., *Medio e Basso Isonzo*, p. 38.
44. Isnenghi and Rochat, *La Grande Guerra*, p. 384.
45. Procacci, 'The Disaster of Caporetto', pp. 151–2.
46. Cited in Walter L. Adamson, 'The Impact of World War I on Italian Political Culture' in Aviel Roshwald and Richard Stites (eds), *European Culture in the Great War: The Arts, Entertainment and Propaganda, 1914–1918* (Cambridge: 1999), p. 323.
47. Emilio Gentile, *Il culto del littorio* (Bari: 1993), pp. 66–74.
48. Malaparte, *Viva Caporetto!*, p. 51.
49. Schivelbusch, *The Culture of Defeat*, pp. 8–10, 30–1.
50. T.C.I., *Medio e Basso Isonzo*, p. 5.
51. 30th Field Artillery Regiment, *Artiglieri del Rubicone*, pp. 58–9.
52. Puccini, *Dal Carso al Piave*, pp. 131–3.
53. C.T.I., *Il Cadore, la Carnia e l'alto Isonzo*, p. 8.
54. Stefano Pivato, *Il Touring Club Italiano* (Bologna: 2006), pp. 103–8. These guides were written with the collaboration of military authorities and prominent ex-servicemen, and closely conformed to the regime's presentation of the war.
55. T.C.I., *Medio e Basso Isonzo*, p. 53.
56. Lucio Fabi, *I Musei della Grande Guerra* (Rovereto: n.d.), p. 92.
57. On the 'religion of the *patria*' see Gentile, *Il culto del littorio*.
58. Indeed many memorials are stylistically very similar to those which commemorate the Risorgimento, in their classical statuary and their use of symbolic elements. Simona Battisti, 'La fabbrica dell'arte: tipologie e modelli' in Vittorio Vidotto, Bruno Tobia and Catherine Brice (eds), *La Memoria Perduta: I monumenti ai caduti della Grande Guerra a Roma e nel Lazio* (Rome: 1998), pp. 39–40; Monteleone and Sarasini, 'I monumenti ai caduti', pp. 632–5.
59. Monteleone and Sarasini, 'I monumenti ai caduti', p. 633; Francesco Bartolini, 'Gloria e Rimpianto. L'evoluzione delle epigrafie' in Vidotto, Bruno Tobia and Catherine Brice (eds), *La Memoria Perduta*, pp. 53–4. On this trend more generally, see Gentile, *Il culto del littorio*, Ch. 1.
60. Bregantin, *Caduti nell'obblio*, pp. 128–9; C.T.I., *Il Cadore, la Carnia e l'alto Isonzo*, p. 249.
61. C.T.I, *Il Cadore, la Carnia e l'alto Isonzo*, p. 249.
62. Ibid.
63. The *via Crucis* in particular is apparently unique.
64. The writer Francesco Perri describes Caporetto as 'our Passion' in his epic poem *La Rapsodia di Caporetto* (Milan: 1919). This work was written in April 1918, and was perhaps a sign of confidence in the rebirth yet to come.
65. Cited in Gentile, *Il culto del littorio*, p. 68.
66. C.T.I., *Sui Campi di Battaglia: Il Trentino, il Pasubio e gli Altipiani*, 1st edn (Milan: 1931), pp. 76–80.
67. Pivato, *Il Touring Club Italiano*, p. 103.
68. Malaparte, *Viva Caporetto!*, p. 197.
69. Bregantin, *Caduti nell'obblio*, pp. 123–4.
70. Gentile, *Il culto del littorio*, pp. 47–8.

71. *Grande Dizionario della Lingua Italiana*, 21 vols (Turin: UTET, 1962); *Grande Dizionario della Lingua Italiana Moderna*, 5 vols (Milan: Garzanti, 1998).
72. U. La Malfa, *Il Caporetto economico* (Milan: 1974). Similarly, 'I saldi estivi fanno flop' in *La Repubblica*, 20 August 2005.
73. For instance, 'Piacenza al capolinea d'Europa', *Gazzetta dello Sport*, 15 December 2004.
74. 'Juventus prese a cannonate', *Gazzetta dello Sport* (28 March 2006). The article describing Juventus's defeat as 'una Caporetto' emphasized that it was the lack of commitment and effort which made this 'unwatchable' defeat especially humiliating.
75. 'Most of them ran away, melting away into the neighbourhood at great speed.' Cited in *Grande Dizionario* (UTET).
76. Procacci, 'The Disaster of Caporetto', p. 157.
77. As used in the article 'Catania in C1, B a 20 Squadre', *Gazzetta dello Sport* (31 July 2003).

5

The Taboos of Defeat: Unmentionable Memories of the Franco-Prussian War in France, 1870–1914

Karine Varley

The French defeat of 1870–1 was widely considered one of the most painful episodes in the nation's history. Having ruled much of Europe scarcely six decades earlier, France was reduced to having to accept territorial mutilation, heavy indemnities, and a humiliating march through the streets of Paris by victorious German forces. If Napoleon III had begun the war seeking to tame a rising Prussia and to boost support for his own ailing regime, it had rapidly degenerated into a rout, culminating in his surrender at Sedan on 2 September 1870. The republican government which took over sought to reinvigorate the defence effort by reviving the spirit of the Revolutionary Wars of the 1790s. Instead of a patriotic rising in defence of the nation, however, there followed only further defeat and capitulation in late January 1871. Protesting betrayal by bourgeois republicans, monarchist factions and the war-weary countryside, national guardsmen and extreme left elements proclaimed a new Paris Commune. A bloody civil war followed, in which around a further 20,000 soldiers and civilians were killed. After such a catalogue of disasters, it would perhaps have been natural for France to have sought to forget *l'année terrible*. Instead, however, the nation appeared to invest more energy in upholding memories of the Franco-Prussian War than it ever had done in fighting it.

Having entered the war confident of victory, defeat came as a profound shock. Many people simply refused to believe they had lost and began calling for *revanche* before the fighting had even ended.[1] But the humiliation ran much deeper than merely being overwhelmed by what had widely been considered a militarily inferior enemy. The collapse of the Second Empire; the military failures of the fledgling Republic; and the intensity of the political, social and cultural polarization that manifested itself in 1871 shattered certainties and resurrected old divisions.

It raised fundamental questions about what it meant to be French, and what the nation had been fighting for. If the war had ended with the surrender of Napoleon III at Sedan, it could have been cast as the downfall of a decaying and decadent regime. But when the republican government picked up the baton and sought to portray the defence in terms of a people's resistance against a reactionary aggressor, it effectively tied the legitimacy of the Republic and the whole mythology of the *levée en masse* to the success of the war effort. Defeat therefore represented more than a military collapse; it came to represent a failure of the Republic and of the nation as well.

The vast and growing body of scholarship on memory emphasizes its subjective, changing and self-serving nature.[2] Thus Pierre Nora observes its constant evolution as it oscillates between the pressures of remembering and forgetting. Indeed, analyses of memory are concerned as much with attempts to forget as with efforts to remember. In psychoanalysis, repression of memory may be regarded either as a response to trauma or discomfort or alternatively as a normal function of the brain's need to create order.[3] For Nietzsche, forgetting is healthy because too much memory can be overwhelming and dangerous, while for Ernest Renan, it is essential for the sake of national unity.[4] Examining the French response to the defeat of 1870–1, however, Robert Gildea argues that instead of coming to terms with the catastrophe, the nation simply took refuge in a kind of denial or 'collective amnesia'.[5] It is not difficult to identify the roots of such a contention. The post-1871 period saw a massive, almost obsessive, outpouring of patriotic representations of the conflict. Motivated by the necessities of national revival, these responses might be regarded as unhealthy and even delusional. Indeed, as Emile Zola commented in 1892: 'There has been in pictures, in literary tales, even in historical annals, a tacit agreement to suppress the failures and the faults, admitting only brilliant acts, the exaltation of patriotism even in the midst of defeat.'[6] In writing *La débâcle*, Zola therefore sought not only to write a 'truthful' account of the war but also to expose the suppression of discomforting aspects of the nation's conduct. Concepts of 'collective amnesia' do not of course necessarily equate with the existence of taboo areas; while the former implies a desire to erase certain memories, the latter suggests merely the avoidance of problematic subjects. An examination of the responses to the defeat of 1870–1 reveals that uncomfortable memories could not be so easily obliterated. In reality, what might have been unmentionable within one political or social environment was often seized upon and manipulated in an opposing other.

With the publication of *La débâcle* in 1892, Zola did not shatter patriotic unity in relation to the war but rather the illusion of patriotic unity. In the aftermath of the defeat, almost every area of cultural activity was mobilized to transform military collapse into moral triumph.[7] The effort was conscious and was considered a patriotic duty. Military artists therefore led a campaign to restore French pride in images of heroic defiance. Thus Alphonse de Neuville's depiction of the final resistance of a handful of soldiers at Bazeilles in *La dernière cartouche* was reproduced in a variety of forms to become a visual shorthand for the patriotic reconfiguration of defeat.[8] Writers of popular literature and poetry such as Paul Déroulède created heroic narratives of struggle against the odds, sacrifice and martyrdom. The prevalence of so many patriotic representations of the war in French culture might appear to suggest that public memories coalesced around a belief that the defeat was more honourable than the German victory. In reality, however, the trend did not signal the emergence of a settled and homogenized public memory that had completely suppressed uncomfortable truths. Patriotic images of a unified national defence reflected not so much public feeling as the aspirations of those who wished to raise the country above its political division towards recovery.

At first sight it might appear that the early Third Republic engaged in a cult of the very institution that had brought the ruin and dismemberment of the nation. Yet the post-1871 glorification of the army is not as paradoxical as it might initially seem. In successfully crushing the Paris Commune, the army not only regained a semblance of competence, but unlike after the June Days of 1848, when its suppression of popular insurrection had made it an object of loathing for many republicans, after 1871 it was hailed by all except the far left as the saviour of honour and the guarantor of order. The experiences of *l'année terrible* thus redefined the relationship between the French nation and its armed forces.[9] Artists flooded the Salons with images of military glory; bookshelves became crammed with military literature; and military education societies sprang up across the country. In a symbolic act of reconciliation with its old foe, the Republic embraced the army, placing it at the centre of Bastille Day celebrations. In 1880, President Jules Grévy distributed new flags to regiments as if to concretize the new era of unity between the army and the Republic. Thereafter, annual popular festivities to mark 14 July were preceded by a military review at Longchamps in Paris and parades of local regiments across the country.

With conservative and moderate republican aspirations for national revival resting so heavily upon the army, its performance in the

Franco-Prussian War needed to be portrayed in an appropriate manner. In the summer of 1871, the monarchist-dominated National Assembly therefore moved to discredit the republican Government of National Defence's handling of the war, attacking the actions of Léon Gambetta, Charles de Freycinet and even Colonel Denfert-Rochereau who led French forces to victory in the siege of Belfort.[10] Viewing irregular forces from the perspective of the Paris Commune, monarchists were joined by moderate republicans in downplaying or even refuting their successes, regarding armed citizens as little more than inciters of indiscipline, insurrection and civil war. If cultural representations of the war sought to portray the improvised character of the national defence in terms that resembled myths of patriotic élan during the Revolutionary Wars, in reality after the armistice of 1871, conservatives and moderate republicans were eager to return to a disciplined professional army.[11]

In other words, for all the insistence upon French glory, the imperatives of patriotic revival did not generate any consensus on the defeat. Indeed, the fall of the Second Empire, the military defeat of the fledgling Republic and the collapse of the Commune produced a vacuum in which the clamour for political power elevated domestic disputes above patriotic considerations. Terrified of anything that might encourage or legitimize far-left militancy, in 1875 the monarchist-dominated Government of Moral Order took the draconian measure of toppling the monument to the defence of Dijon. Asserting that the 'revolutionary nature' of Paul Cabet's statue, which featured a Phrygian cap, rendered it an image of republicanism that was 'subversive and aspiring to disorder', the government sought to quash any suggestion that citizens in arms had brought glory to the war effort.[12] Having judged the defence of Dijon to be futile against such overwhelming enemy forces, on 28 October 1870, the army had abandoned it, withdrawn and ordered the disarmament of the city's National Guard. Refusing to surrender, however, national guardsmen, volunteers and other irregular combatants demanded resistance.[13] Against an army of 10,000 men, they managed to hold off the assault for several hours, securing terms of surrender usually only accorded to armies that had endured lengthy sieges. For the Government of Moral Order, the contribution such a story of heroism might make to the national revival mattered less than the political implications of the popular resistance. Haunted by the spectre of the Paris Commune, Interior Minister Buffet calculated that the dangers of appearing to sanction popular republicanism were sufficiently grave as to demand the immediate removal of the war memorial by force.[14] Moderate republicans also chose to play down the role of irregular soldiers in a bid to limit potentially damaging associations with the patriotic republicanism of the Commune. Whereas Gambetta had sought to revive the spirit that had supposedly animated the nation during the Revolutionary Wars, calling for a *levée en masse* in September 1870, the disasters of *l'année terrible* compelled him to reflect that the emphasis upon irregular forces had been a mistake. Indeed, as Gerd Krumeich observes, in late 1871 Gambetta and Freycinet abandoned their ideas of a nation in arms in favour of returning to a more conservative model of the army.[15]

With so much effort invested in presenting the regular army as the saviour of French honour, the debacle at Sedan was doubly embarrassing. For the most part, cultural representations of the war concentrated upon the period under the republican Government of National Defence, but Sedan represented much more than a military disaster. For Bonapartists, Sedan was inseparable from recollections of the revolution of 4 September and the 'treachery' of republicans who brought revolution while the country was resisting invasion.[16] For republicans, Sedan represented the betrayal of Napoleon III, whom they claimed had sacrificed France in the interest of protecting his own dynasty. For the Catholic Church, Sedan was divine punishment for the sins committed by the Second Empire. Such was the shadow cast by Sedan that even the heroic cavalry charge at Floing which earned the admiration of Wilhelm I was not widely celebrated in art or literature, and it took until 1910 for a monument to be erected in its memory.

In August 1877, the Ministry of Public Instruction and Fine Arts commissioned Edouard Armand-Dumaresq to produce a painting of the moment when Marshal MacMahon was wounded in the battle of Sedan.[17] Intended for display in the museum of Versailles, it told the story of how the commander of the Army of Châlons was put out of the action after being hit by fragments from a German shell as he rode out to assess the situation in Bazeilles. Despite being partially responsible for the army's disastrous performance at Sedan and Froeschwiller, MacMahon had largely escaped blame for the defeat by re-establishing his reputation commanding the Army of Versailles against the forces of the Paris Commune. With monarchist credentials and a record of loyalty to the state, he was chosen to replace Adolphe Thiers as president in 1873. The initial report on Armand-Dumaresq's work found that if the painting lacked eloquence, it did at least portray the wounded MacMahon maintaining his courage and refusing to be beaten.[18] The verdict from MacMahon's family in 1879 was, however, considerably more critical. According to a report sent to the Director General of the Ministry of Fine Arts, all those who had seen the painting had concluded that Armand-Dumaresq had got it 'absolutely wrong', having 'effectively plunged into the memory of a wound which [. . .] became the point of departure for one of the most painful national catastrophes in the annals of history'.[19] To be sure, while the depiction of the wounding of MacMahon might have offered a flattering portrayal of his war record, the incident did not occur in the midst of battle. Instead, the viewer was presented with an image of a soldier's physical fragility as the painting drew attention to the weakness of France's defenders. Rather than obscuring the defeat behind a narrative of heroism, it seemed a blunt reminder of reality. The official who examined the painting therefore recommended that in future the administration be more selective, commissioning only those works of art which 'retrace our past glories'.[20] The advice was heeded and if paintings of the Franco-Prussian War represented one-third of the military art purchased by the state, it was virtually absent from the museum of Versailles.[21]

The enduring discomfort with memories of Sedan was also manifested in the debates over the construction of a war memorial for the centre of the town. The stimulus for the project lay in local perceptions that the town had been unduly stigmatized and tainted with memories of national humiliation.[22] Seeking to reshape recollections of the battle of Sedan in terms of the heroism and sacrifices of the French army, the town's monument committee presented itself as the non-partisan incarnation of local patriotism. Such naïve ambitions clashed with ministerial concerns about resurrecting memories of such a divisive and painful episode in the nation's recent history, however. The former mayor of Sedan offered the prime minister the honorary presidency of the committee, only for Freycinet to decline, refusing to allow subscription lists to be circulated within the army, and denying the prefect and sub-prefect of the Ardennes permission to act as patrons.[23] Many senior ministers also chose not to be associated with the initiative. They did not object to the monument *per se* so much as the proposal to give it a national character. With one eye on diplomatic relations with Germany and another on domestic politics, they argued that it should be limited to a purely regional project. Indeed, so damaging were its connotations that 20 years after it was the scene of military and political debacle, Sedan still represented a most awkward and uncomfortable subject for the Republic.

Ministerial efforts to stifle memories of Sedan and to limit its presence in national political discourse prompted the deputy and former mayor, Auguste Philippoteaux, to claim that it would be 'the last town in France forbidden to erect a monument in memory of the soldiers who died within its walls'.[24] He invited the committee to prove to the government that the memorial would arouse patriotic sentiment rather than 'painful emotions' and that it would transform memories of 1870 from humiliation to honour.[25] The completion of the monument brought no change in government thinking, however. In January 1896 the prefect of the Seine sought to prevent its public exhibition on the grounds that it would revive too many painful memories for Parisians.[26] Moreover, after a little initial hesitation, ministers in Méline's government decided to give the unveiling ceremony which was due to take place on 7 August 1897 a wide berth.[27] If the patriotic reconfiguration of the war had elsewhere turned humiliating defeats into moral victories, the stigma attached to Sedan meant that it could not be presented as anything other than a painful episode in French history. Indeed, as *La République Française* and the moderate republican *Les Ardennes* commented, the committee's desire to honour the heroism of the army at Sedan would not only appear inappropriate but would lack credibility as well.[28] The collapse of the Second Empire might have made Sedan a cause for celebration for its opponents, but to have taken political satisfaction from a national disaster would have been to aggravate the divisive legacy of the war. As Gambetta explained in 1872, even celebrating the anniversary of the proclamation of the Republic on 4 September was problematic, 'because it was followed by events which the whole of France would like to erase from our history'.[29] Regardless of how memories of Sedan might have been shaped to further political or other agendas, ultimately the national disaster benefited no one.

Diplomatic relations with Germany also problematized remembrance of the Franco-Prussian War. In the decade following the defeat, tension reigned between the former enemies, manifesting itself most dramatically in the 'war in sight' crisis of 1875. Ever sensitive to potential repercussions from Germany, at several points over the course of the 1870s, the government intervened to restrict public expression of bellicose or *revanchist* sentiment. The terms of the Treaty of Frankfurt were such that even after the evacuation of occupying forces, France remained obliged to uphold German interests. The Treaty essentially defined the relationship between the two nations, tying them together with the obligation that France accord Germany most favoured nation status in matters of commerce. Under the terms of Article 16, moreover, each power agreed to preserve the resting-places of each other's fallen soldiers. French *lieux de mémoire* were thus also German *lieux de mémoire*. With the annexation of Alsace-Lorraine representing a source of long-term French bitterness, the German government and press were acutely sensitive to any signs of *revanchism* across the border.

At two of the most sensitive sites relating to the war, Mars-la-Tour and Sedan, officials intervened to curtail anything that might have been negatively construed in Germany. Despite the insistence of Sedan's mayor that his town merited an ossuary commensurate with its historical significance and the painfulness of the memories it evoked, the Interior Ministry declined, citing diplomatic concerns.[30] Constructed by the state as part of the implementation of Article 16 of the Treaty of Frankfurt, the ossuary had to contain the remains of French and German soldiers. With Sedan representing a defining point in the road to German unification, however, any suggestion that the resting-place of the fallen was being used to launch a French counter-memory risked being interpreted as an act of provocation. At the behest of officials, municipal authorities in Sedan therefore agreed to keep the ossuary modest, austere and neutral in style, and to construct a separate memorial bestowing glory upon the defeated.[31] In Mars-la-Tour on the edge of Lorraine's new border with Germany, meanwhile, plans to unveil the memorial in a major ceremony in 1875 had to be postponed until diplomatic tensions with Germany had eased.[32]

Ministerial intervention in the Universal Exposition of 1878 was rather more chaotic and controversial. Having initially said that it would not attend, the German government's last-minute change of mind prompted an emergency Cabinet proclamation that 'works of art whose subject relates to the events of 1870–1 shall not be admitted to the Universal Exposition or the Annual Exposition of 1878'.[33] Even exhibits making only indirect allusion to the war were excluded.[34] Thus Eugène Bellangé's *Au drapeau*, which was based upon the principal group in his earlier painting, *91e de ligne à Solferino*, fell foul of the ban, as did several of his landscapes.[35] The wide-ranging nature of the ban for which no compensation was offered left many artists feeling that they were having to shoulder the burden for a diplomatic crisis not of their making.[36] While the incident had a precedent in the removal of works of art relating to the war from the Salon of 1872, it was the first time such measures had been taken since the departure of occupying forces.[37] The reality was, however, that if government intervention represented the most draconian restriction upon cultural expression, artists were nonetheless compelled to operate in an environment where representations of the war were constantly constrained by political, diplomatic and public sensitivities.

Ultimately, memories of the Franco-Prussian War revolved around the dilemmas of presenting the defeat in such a manner as to derive political advantage from it or to diminish its significance. Before 1870, military art had largely focused upon great battle scenes, depicting victory rather than defeat. In the period thereafter, however, the expansion of military service produced 'armchair generals' and a new public interest in all areas of army life. As the art critic Jules Richard commented in 1888, the defeat transformed all citizens into soldiers, having as their duty the revival of the nation.[38] With the army engaged in extending France's colonial possessions during the 1880s, military conflict became a part of daily life and discourse. Yet if war was no longer portrayed in such glorious terms, representations of the defeat of 1870 were scarcely embedded in reality either. Indeed, when confronted with a disintegrating army in Alfred Roll's *La guerre – marche en avant*, critics remarked that such images were demoralizing and ought not to be presented to the public.[39] Instead of war-related paintings being viewed in terms of their artistic merit, they were assessed according to their perceived national utility. Critics and audiences preferred not to see the full magnitude of the defeat and at the very least to see it depicted in consoling terms. Thus Eugène Médard's *Buzenval (1870)* portraying the final sortie from besieged Paris on 18 January 1871 seemed more appealing for showing soldiers' struggle to save French honour in the face of certain defeat.[40]

Even before the Franco-Prussian War, military art had been changing as a genre, reacting against the old bellicose style of official battle painting in favour of a focus upon smaller-scale depictions of the hardships faced by ordinary soldiers.[41] Death on the battlefield was not a taboo subject so long as sacrifices were rendered glorious; for as Jules Richard commented, the public 'does not like to be told the truth too cruelly'.[42] Here too, however, recent developments had brought changes in attitude and approach. In the wake of the American Civil War when corpses on the battlefield had been photographed for public exhibition, artists became more willing to depict the dead, albeit still within a restrictive framework of public acceptability. Auguste Lançon may have portrayed the burial of the dead at Champigny in a brutally stark etching, but as John Milner observes, the non-specific setting of the image gave it a more universal message of human frailty and mortality.[43] And while the likes of Zola, Maupassant, Mirbeau and Huysmans criticized the obligations imposed upon military artists, ultimately those who did choose to depict the war in realistic terms did so in the full knowledge that they faced accusations of patriotic betrayal and rejection by Salon juries.[44] Many therefore simply avoided engaging in such a painful and problematic subject.

Revanche may never have been seriously contemplated within government, but for much of the country, preparing for future conflict helped to ease the pain of defeat by perpetuating the notion that the war had not been definitively lost. The post-1871 cult of the army was just one aspect of a wider trend towards not just the glorification of patriotic self-sacrifice but its elevation into the ultimate and indeed only true act of devotion to the nation. Death therefore had to be configured in heroic terms, especially in republican discourse seeking to compete with deeply entrenched Christian ideas. For grieving families and communities, it also had to be presented in a comforting manner, while for veterans the sanctity of their fallen comrades remained a priority. There were thus important political and social imperatives behind the sanitization of death in combat. That such considerations should have been disregarded in the construction of the ossuary at Bazeilles and in Zola's *La débâcle* rendered their impact all the greater.

Ever since the earliest efforts to grant fallen soldiers a permanent resting-place, questions have been raised about how war should be represented. Indeed, carefully maintained cemeteries and beautifully sculptured monuments might be considered fitting tributes to those who have sacrificed their lives, but they might also appear to sanitize war, effectively masking its brutality.[45] In the devastated village of Bazeilles, however, the ossuary containing the remains of all those who had died in the battle, including civilians, was designed to produce the opposite effect. Visitors could enter and view for themselves the skeletons of over two thousand victims separated into two piles according to nationality. The resulting effect was devastatingly stark and horrific. Those who recorded their impressions described their revulsion at seeing clothing still shrouding some of the bones, a foot still in its shoe and fingers still wearing wedding rings.[46] With no surviving records on the construction of the ossuary, one can only surmise that the reason it was opened to the public was to reveal the remains of 2059 victims of war in their full horror.[47] The battle for Bazeilles between 31 August and 1 September 1870 became the subject of international scandal after the village was destroyed in fires that French witnesses claimed had been deliberately started by Bavarian soldiers.[48] In the chaos of war, journalists reported that the civilian death toll had reached into thousands, causing 'one of the most deplorable incidents of modern warfare'.[49] The ossuary reversed usual practices of sanitizing death in combat; its exterior may have presented an attractive, decorative image, but the victims inside were recognizably individuals, the personal items still attached to some skeletons bearing witness to the lives they had once led. It would be an overstatement to describe the ossuary as an expression of pacifism, for such sentiments were rare at a time of patriotic reaction to the defeat, but the ossuary unquestionably sought to testify to the destructiveness of modern warfare.

Aside from its highly critical depiction of the conduct of French soldiers and civilians during the war, one of the most striking aspects of Zola's *La débâcle* is the realism of the battle scenes. As a self-proclaimed naturalist writer, Zola approached the events of *l'année terrible* with an interest in social observation and a desire to present the truth. Claiming to have based his work on extensive interviews with eyewitnesses as well as documentary evidence, Zola's descriptions of the devastating effects of modern warfare evidently aimed to shock readers.[50] Almost 20 years earlier, M. L. Gagneur's *Chair à canon* had provoked uproar with its brutal realism, while reactions against violent images in military art had kept the debate about representations of the war alive.[51] For Zola, war was not a moment of glory but men choking on their own blood, others with their entrails blown out by enemy artillery, bullets shattering soldiers' skulls or the final agonizing convulsions of death.[52] Heroism was futile in the face of annihilation by German breech-loading rifles. The fate that awaited men after their death was scarcely more glorious; Zola described the stench of rotting flesh and bodies being unceremoniously shovelled into refuse carts the morning after battle.[53] As a well-established writer, Zola was fully aware of the impact his book would have upon critics and the wider public. Despite his stated intention to create a work of truth, the ferocity of his assault upon post-war myths was such that it exposed the ways in which patriotic discourse had created its own taboo subjects. If claims about desertion, indiscipline and peasant indifference to the war effort were well-known and had already been debated and contested in a variety of milieux, the graphic and detailed nature of Zola's portrayal of modern warfare was new. A handful of artists and writers had previously attempted to present more realistic depictions of the war, but none had done so with such aggressive irreverence.

Although the likes of Philippe Gille, Anatole France and Paul Alexis praised the rare realism of *La débâcle's* battle scenes, most critics maintained that Zola had not only neglected his patriotic duties but had gone so far as to harm the national revival.[54] Assessing the book in terms of what they considered its function ought to have been, Zola's critics suggested that it amounted to little more than a betrayal. Thus General Morel complained that 'Zola could have written a useful book, celebrating, patriotic, encouraging for future generations, instructive for future governments', but that instead he had chosen to write one that was 'scandalous and immoral'.[55] Jules Arnaud went still further, accusing

Zola of seeking to enrich himself by 'poisoning France'.[56] Rather than seeing realism in the scenes of battle, he found soldiers' deaths to have been trivialized and even rendered ridiculous.[57] For his part, Zola maintained that he had deliberately rejected traditional depictions of soldiers as one-dimensional heroes feeling nothing but bravery on the grounds that 'it has almost seemed to be a crime of *lèse patrie* to suppose for a moment that soldiers can ever have fear, and that a man with his miseries may be on the field what he is everywhere else.'[58] The desire to portray soldiers' emotions on the battlefield might have been driven by the imperatives of the naturalist literary genre but it also resulted from a necessary confrontation with the realities of modern warfare. The destructiveness of new guns and cannon raised questions about the conduct of future war. If post-war patriotic discourse sought to celebrate French resilience in the face of the overwhelming strength of the enemy, it was out of a tacit realization that heroism alone was no longer a credible response to superior firepower. Indeed, the experiences of *l'année terrible* forced a confrontation with myths of patriotic élan dating from the Revolutionary Wars. Technological advances had irrevocably altered the nature of warfare, but Zola's subsequent conclusions offered little hope for the recovery of national pride, for 'what we buried at Sedan was the legend of our bellicose spirit, that legend of French troops going off to conquer neighbouring kingdoms for nothing, for pleasure. With new weapons, war has become something terrible'.[59] With patriotic revival and the idea of revenge foremost in nationalist discourse, it is hardly surprising that so many rushed to refute Zola's claims.

Ultimately, however, the greatest taboo subject of *l'année terrible* was the Paris Commune. All but the far left made concerted efforts to remove the Commune from memories of the Franco-Prussian War, not just because it was an episode they would have preferred to forget, but also because it represented an enduring threat to national unity. Because the uprising had occurred in defiance of the surrender, concepts of patriotism had to be wrested from the Commune such that notions of *guerre à outrance* were resolutely rejected as damaging to the nation. The spectre of a revival of the Commune haunted all but its sympathizers throughout the 1870s; the right viewed rising republican support in the polls as a sign of resurgent left extremism, while many moderate republicans desperately sought to dissociate themselves from its damaging associations. In literature, art and culture more generally, the Commune was largely ignored for around two decades in favour of more unifying historical subjects.[60] Thereafter, its suppression was presented by the likes of Zola and the Margueritte brothers as the beginnings of national rebirth, the necessary purging of corruptive elements.[61] The divisiveness of the Commune was thus submerged beneath narratives of unity as few within the cultural mainstream wished to bestow glory upon those who were widely considered responsible for causing widespread death and destruction.

For all the efforts to exorcize the Commune from memories of *l'année terrible*, the Communards themselves could not be so easily dismissed. Their return to France following the amnesty of 1880 caused widespread fear among their political enemies. Even the Communard dead seemed to represent a persistent threat to political and social stability. Those who had died in the fighting of April and May 1871 were hastily buried in secret mass graves by a government worried about the political ramifications of their deaths. Successive administrations, whether conservative or moderate republican, refused to sanction funeral ceremonies or the construction of memorials to the Commune, wishing to avoid creating sites of pilgrimage for the far left. Thus the unprecedented efforts to remember the Franco-Prussian War were paralleled by even greater efforts to forget the Paris Commune. The problem was, however, that the two could not be so easily separated; remembering the Franco-Prussian War often also involved remembering the Commune.

With *l'année terrible* remaining a painfully acute presence in national consciousness during the 1870s and 1880s, memories of the Commune periodically resurfaced and disrupted memories of the Franco-Prussian War. Because the shadow of the Commune loomed so large over the capital, political expediency dictated that the *conseil général de la Seine* choose a relatively anodyne statue for the monument to the defence of Paris. The jury thus rejected Rodin's rather injudicious entry to the competition which featured a screaming allegory of Paris wearing a Phrygian cap in favour of Ernest Barrias' more sober representation of the capital dressed the greatcoat of a national guardsman.[62] The unveiling ceremony in August 1883 was carefully stage-managed to limit the intrusion of awkward memories; everyone except the radical president of the *conseil général de la Seine*, Barthélemy Forest, was forbidden from addressing the crowds. Organizers could not prevent participants from expressing their grievances about the events of *l'année terrible*, however. Signalling their continuing hostility towards the Commune, military bands marched in silence past the municipal council of Paris. If most of the soldiers who had fought for the Army of Versailles against the forces of the Commune had not been ideologically driven fanatics, there nonetheless remained hostility within the regular army against those exponents of left extremism.[63] In the midst of the parade of the workers' associations and free-thinking societies, a man unfurled a red flag and brandished it at Interior Minister Waldeck-Rousseau. During the civil war, the red flag had served as a symbol of Communard patriotism in opposition to the tricolour's association with surrender and Versailles. As workers shouted, 'Long live the amnesty! Long live the social revolution!', they revealed the impossibility of any separation between memories of the Franco-Prussian War and the Commune in the capital.[64] In celebrating Parisian resistance under siege, it was impossible to ignore the fact that the Commune had arisen in the name of *résistance à outrance*.

On one level, discomfort with memories of the Commune was a consequence of moderate republican awkwardness over potentially damaging associations with the extreme left. It was also a measure of the divisions within the left between those advocating a republic of social reform and those in favour of bourgeois republicanism. On another level, however, the Commune presented its detractors with a paradox: discrediting it meant having to argue in favour of surrender as an act of patriotism. For those who had supported Gambetta's calls for *guerre à outrance* and a *levée en masse* in the manner of 1793, it was especially problematic. Indeed, such was the case for Paul Déroulède, the nationalist poet and leader of the *Ligue des Patriotes*, who became the most vociferous campaigner for *revanche* by arguing that the war with Germany would not be over until the restitution of Alsace-Lorraine. Déroulède resolutely rejected the Commune as a legitimate expression of patriotism, condemning its political and social divisiveness as harmful to national unity.[65] Having served in the Army of Versailles, however, he could later be taunted by his political enemies for having opposed supporters of the Commune as they rose up against the peace terms which brought the dismemberment of the nation.[66] Seeking to confront and embarrass Déroulède, members of local freethinking societies interrupted the war commemorations at Buzenval in January 1886 with red flags and loud acclamations of the Commune.[67] The battle marked the final attempt to break out of the besieged capital, but in the period thereafter, many came to regard it with the suspicion that General Trochu had only sent the National Guard into action as a means of killing off some of its most radical elements.[68] By introducing red flags in opposition to the tricolour, the protestors effectively challenged the hegemony of moderate republican and right-wing discourses over memories of the Franco-Prussian War, seeking to re-establish the Commune as a legitimate expression of defiance against defeat.

Why were memories of the defeat of 1870–1 so problematic? France had, after all, suffered such humiliations before, and had re-emerged, revived and reinvigorated. In part, the difficulty arose from the fact that whereas the France of Napoleon I had been defeated by an alliance of nations, the France of Napoleon III and the Third Republic was defeated by only one. Perhaps even more importantly, the experiences of *l'année terrible* appeared to shatter the grandeur of France's past: the glories of the First Empire were demolished by the routing of Napoleon III, while myths of the nation in arms of 1793 were exploded by the failings of Gambetta's attempted *levée en masse*.[69] Was there anything specifically French in the problematic confrontation with defeat after 1871? Success or failure in war has often been considered a measure of national strength, but such views were particularly prevalent in late nineteenth-century Europe. Influenced by discourse on race and degeneration, many commentators in France and across the continent viewed the French disaster as a sign of national decline.[70] For France, however, the trauma of defeat was particularly acute because it revived questions about the revolution of 1789 and the character of the nation. Indeed, a revived Catholic Church led the way in arguing that the disasters which had befallen the nation were God's punishment for France's deviation from its divinely ordained path. Enjoying a revival in strength and political support from monarchists in government in the 1870s, the Church advanced a rival concept of the French nation rooted in Catholicism. The failure of the Government of National Defence to mobilize the kind of national zeal that had supposedly expelled the enemy in the Revolutionary Wars represented a major blow to republican assertions of legitimacy. Republican claims to be the only true representatives of the nation were simply not borne out by the reactions of war-weary provincial populations; in response to Gambetta's calls for *guerre à outrance*, the country voted overwhelmingly for peace on 8 February 1871. Balanced precariously between adversarial allies, the Government of National Defence was unable to contain the excesses of the extreme left. The rising of the Paris Commune in opposition to the surrender crystallized the problematic political equations for the bourgeois Republic as it sought to distance itself from notions of social upheaval, violent revolution and civil war.

The fact that patriotic discourse sought to conceal and even deny divergent perspectives on the defeat is scarcely surprising in view of the hybrid nature of the conflict. Beginning as an attempt to boost support for the Second Empire, it then became a republican-led war of national defence; and then in March 1871, it turned into civil war. Professional soldiers, mobilized National Guards, volunteers and civilians, all experienced the war in very different terms. Many of those conscripted in the late 1860s were of peasant origin, armed with an essentially conservative outlook, whereas many of the *mobiles* held republican views. Volunteers entered the war with other objectives: *francs-tireurs* under Garibaldi were inspired by revolutionary and anti-clerical convictions, whereas Papal Zouaves fought behind banners of the Sacré Coeur. The war may have witnessed a fragile political truce, but it remained the case that many of those who fought it did so for very different reasons.

Notes

1. S. Audoin-Rouzeau, 'French Public Opinion in 1870–71 and the Emergence of Total War' in S. Förster and J. Nagler (eds), *On the Road to Total War: The American Civil War and the German Wars of Unification, 1861–1871* (Cambridge: Cambridge University Press, 1997), p. 407.
2. On concepts of collective memories, see M. Halbwachs, *On Collective Memory* (Chicago, London: University of Chicago, 1992); A. Confino, 'Collective Memory and Cultural History: Problem of Method', *American Historical Review*, 102 (1997), 1393–4; P. Fritzsche, 'The Case of Modern Memory', *Journal of Modern History*, 73 (2001), 87–117; J. Fentress, and C. Wickham, *Social Memory* (Oxford: Blackwell, 1992), p. 134.
3. M. Wehner, 'Typologies of Memory and Forgetting among the Expatriates of Raboul', *Journal of Pacific History*, 37 (2002), 72.
4. See E. Renan, *Qu'est-ce qu'une nation? Conférence faite en Sorbonne, le 11 mars 1882* (Paris: C. Lévy, 1882).
5. R. Gildea, *The Past in French History* (New Haven: Yale University Press, 1994), pp. 119–21.
6. *The Times*, 11 October 1892.
7. C. Digeon, *La crise allemande de la pensée française* (Paris: Presses universitaires de France, 1959); F. Roth, *La guerre de 70* (Paris: Fayard, 1990).
8. See F. Robichon, *L'armée française vue par les peintres 1870–1914* (Paris: Herscher, 2000).
9. J. Horne, 'Defining the Enemy: War, Law and the *levée en masse* from 1870 to 1945' in D. Moran and A. Waldron (eds), *The People in Arms: Military Myth and National Mobilization since the French Revolution* (Cambridge: Cambridge University Press, 2003), p. 110.
10. *Enquête parlementaire sur les actes du gouvernement de la défense nationale. Tome 1: Dépositions des temois* (Versailles: Cerf et Fils, 1872), pp. 571, 207; Municipal Council of Belfort minutes, 7 June 1878, Archives Municipales de Belfort 1M 31; see also W. Serman, 'Denfert-Rochereau et la discipline dans l'armée française entre 1895 et 1874', *Revue d'histoire moderne et contemporaine*, 20 (1973), 95–103.
11. See R. Girardet, *La société militaire dans la France contemporaine 1815–1939* (Paris: Plon, 1953); R. D. Challener, *The French Theory of the Nation in Arms 1866–1939* (New York: Columbia University Press, 1955).

12. Letter from the Prefect of the Côte-d'Or to Mayor Enfert, 21 October 1875, Archives Municipales de Dijon, 1M 16; *Le Siècle*, 28 October 1875.
13. P. A. Dormoy, *Les trois batailles de Dijon 30 octobre – 26 novembre, 21 janvier* (Paris: E. Dubois, 1894), p. 61.
14. *Le Siècle*, 28 October 1875.
15. G. Krumeich, 'The Myth of Gambetta and the "People's War" in Germany and France, 1871–1914' in Förster and Nagler (eds), *On the Road*, pp. 650–1.
16. J. El Gammal, 'La guerre de 1870–1871 dans la mémoire des droites' in J.-F. Sirinelli (ed.), *Histoire des droites en France. Vol. 2: Cultures* (Paris: Gallimard, 1992), p. 475.
17. Memorandum from the Ministry of Public Instruction and Fine Arts, 13 August 1877, Archives Nationales (hereafter AN) F21 191.
18. Internal note to the Ministry of Public Instruction and Fine Arts, 10 January 1878, AN F21 191.
19. Report to the Director General of Fine Arts, 1 February 1879, AN F21 191.
20. Ibid.
21. F. Robichon, *La peinture militaire française de 1871 à 1914* (Paris: Association des amis d'Edouard Detaille, Bernard Giovananeli Editeur, 1998), p. 62.
22. Meeting of the provisional monument committee, 22 November 1890, Archives de la Ville de Sedan (hereafter AVS) H218.
23. Minutes of the provisional monument committee, 27 December 1890, AVS H218.
24. Ibid.
25. Ibid.
26. *Le Figaro*, 25 January 1896.
27. Letter from Boucher, 22 July 1897, AVS H218.
28. *La République Française*, 31 July 1897; *Les Ardennes*, 30 July 1897.
29. *Le Siècle*, 6 September 1872.
30. Letter from the architect Racine to the Prefect of the Ardennes, 20 September 1876; letter from Morin Maisson to the Sub-Prefect of the Ardennes, 27 February 1876, AN F9 1358.
31. Letter from the Assistant Mayor to the Prefect of the Ardennes, 16 October 1877; Letter from Racine to the Prefect 8 December 1877, AN F9 1358.
32. Letter from the Prefect of Meurthe et Moselle to the President of the Conseil Général, 15 May 1875; Letter from the Prefect of Meurthe et Moselle to the Interior Minister, 31 October 1875; Urgent letter from the Prefect to the Interior Minister, 1 November 1875, Archives Départementales de Meurthe et Moselle 1M 670.
33. Notice from the Ministry of Public Instruction and Fine Arts, 20 April 1878, AN F21 524.
34. Letter from the Ministry of Public Instruction and Fine Arts, no date, AN F21 524.
35. Letter from Eugène Bellangé fils, 9 May 1878, AN F21 524.
36. Letter from Charles Castellani to the Ministry of Public Instruction and Fine Arts, 14 May 1878, AN F21 524.
37. J. Milner, *Art, War and Revolution in France 1870–1871: Myth, Reportage and Reality* (New Haven: Yale University Press, 2000), p. 195.
38. J. Richard, *Le Salon Militaire de 1886 (Première année)* (Paris: Jules Moutonnet Editeur, 1886), pp. 4–7

39. J. Richard, *Le Salon Militaire de 1887 (Deuxième année)* (Paris: Alphonse Piaget Editeur, 1887), pp. 26–8; F. Robichon, 'Representing the 1870–1871 War, or the Impossible *Revanche*' in N. McWilliam and J. Hargrove (eds), *Nationalism and French Visual Culture 1870–1914* (New Haven: Yale University Press, 2005), p. 88.
40. Richard, *Salon Militaire de 1886*, pp. 70–1.
41. Ibid., 4–11; see also Robichon, *La peinture militaire*, 86.
42. Ibid., 71
43. Milner, *Art, War and Revolution*, p. 100.
44. Robichon, *La peinture militaire*, p. 70.
45. D. W. Lloyd, *Battlefield Tourism: Pilgrimage and Commemoration of the Great War in Britain, Australia and Canada, 1919–1939* (Oxford: Berg, 1998), pp. 112–29.
46. J. Bourgerie, *Bazeilles combats, incendies, massacres* (Tours: P. Bousrez, 1897), p. 108; A. Meyrac, *Géographie illustrée des Ardennes* (Charleville: Edouard Jolly, 1900).
47. I am indebted to the guard of the ossuary for suggesting this to me.
48. M. R. Stoneman, 'The Bavarian Army and French Civilians in the War of 1870–1871: A Cultural Interpretation', *War in History*, 8 (2001), 276.
49. *The Times*, 16 September 1870.
50. E. Zola, 'Retour du voyage', *Le Figaro*, 10 October 1892.
51. Digeon, *La Crise allemande*, p. 51.
52. E. Zola, *The Debacle 1870–1* (London: Penguin, 1972), pp. 257, 264, 269.
53. Ibid., 343.
54. *Le Figaro*, 21 June 1892; *Le Temps*, 26 June 1892; *Gil Blas*, 21 June 1892.
55. L. Morel, *A propos de 'La Débâcle'* (Paris: H. Charles-Lavauzelle, 1893), p. 39.
56. J. Arnaud, *La Débâcle de M. Zola* (Nîmes: Imprimerie de Lessertisseux, 1892), pp. 20, 17.
57. Ibid., pp. 12, 17.
58. *The Times*, 11 October 1892.
59. *Le Figaro*, 1 September 1891.
60. R. Tombs, *The Paris Commune, 1871* (London: Longman, 1999), p. 193.
61. Zola, *The Debacle*, pp. 505–9; Paul and Victor Margueritte, *Une époque. Vol. 4: La Commune* (Paris: Plon-Nourrit, 1904), pp. 625–6.
62. Minutes of the Conseil général de la Seine, 30 November 1878 and 29 April 1879, Archives de Paris VR 161; Letter from the Prefect of the Seine, 9 January 1879, AN F1cI 172; G. Coquiot, *Rodin à l'hôtel de Biron et à Meudon* (Paris: Ollendorff, 1917), p. 107.
63. R. Tombs, *The War against Paris 1871* (Cambridge: Cambridge University Press, 1981), pp. 105–8.
64. Patrick Chamouard, 'Un après-midi d'été 1883: l'inauguration de la statue' in Georges Weill (ed.) *La Perspective de la défense dans l'art et l'histoire* (Nanterre: Archives départementales des Hauts-de-Seine, 1983), p. 165; *Le Petit Parisien*, 14 August 1883; *L'Intransigeant*, 14 August 1883.
65. B. Joly, *Déroulède: L'Inventeur du nationalisme* (Paris: Perrin, 1998), p. 30.
66. *L'Intransigeant*, 4 March 1885.
67. *Le Drapeau*, 23 January 1886, 6 February 1886.
68. A. Horne, *The Fall of Paris: The Siege and the Commune 1870–1* (London: Pan, 1965), p. 234.

69. E. Renan, *La réforme intellectuelle et morale. Oeuvres Complètes Tome 1* (Paris: Calmann-Lévy, 1947), pp. 334–5.
70. See D. Pick, *Faces of Degeneration: A European Disorder c. 1848–c. 1918* (Cambridge: Cambridge University Press, 1989); D. Pick, *War Machine: The Rationalisation of Slaughter in the Modern Age* (New Haven and London: Yale University Press, 1993).

6

Religious War, German War, Total War: The Shadow of the Thirty Years' War on German War Making in the Twentieth Century

Kevin Cramer

The German understanding of war as religious war scourged Europe in the first half of the twentieth century. This understanding was, in large measure, a legacy of nineteenth-century Germany's working through the meaning of the Thirty Years' War, the horrifically destructive conflict ignited in 1618 by the collapse of the fragile peace between the Protestant princes of the Holy Roman Empire and the Catholic Habsburg emperors. Driven by all or nothing strategies and waged with maximal violence in pursuit of radically transformative ends, German war making in the two world wars seemed to fundamentally contradict Clausewitz's pragmatic dictum that 'war cannot be divorced from political life'.[1] In the belief that he was defining (and limiting) war as a legitimate instrument of rationally determined policy, interpretations of Clausewitz's theories on war have conventionally buttressed analyses of the rise of Frederician Prussia as a model of the enlightened absolutist state. It is sometimes forgotten, however, that Clausewitz saw in the advent of revolutionary France's ideologically mobilized citizen armies the return of war 'in all its elemental fury', a regression to the 'war of extermination', or 'absolute war'.[2] Clausewitz's iconic status within German military culture has often been cited as a partial explanation for the willingness of that nation's leadership to contemplate war as a panacea for crises, foreign and domestic. But unless historians recognize that Clausewitz also understood the German historical experience of war as a demonstration of the possibility (and potential) of a war of annihilation, an important and distinguishing characteristic of German war making will remain obscured.

Clausewitz's apprehension reflected as well the historical interpretations and popular memory of the loss of life and material devastation of

Thirty Years' War which influenced an emergent German national identity as it coalesced around Protestant nationalism in the early nineteenth century. Beginning in the period of the Napoleonic Wars, a burgeoning historical literature on the war, comprising histories, plays, novels, poems, and rediscovered memoirs, took up a broad rediscovery and reassessment of the Thirty Years' War as the key to understanding Germany's present and divining its future.[3] In a century that witnessed the birth of modern German nationalism and the triumphal unification of Germany as a powerful nation state, the Germans' obsession with the most destructive and divisive war in their history is remarkable, given that the story of the Thirty Years' War was essentially a narrative of disunity, destruction, and defeat. The explanation of this phenomenon can be found in the way Protestant nationalism asserted the eschatological nature of the conflict between Protestant and Catholic narratives that relied heavily on the Old Testament themes of divine judgement, enslavement and captivity, sacrifice and redemption, and, more particularly, the idea of the covenant between God and his chosen people (reified as the nation).[4] In this conception of nationhood it is *religious war* which establishes the rule of the elect. This is the conception of war, as it emerged from the nineteenth-century histories of the Thirty Years' War, that helps explain the extremes of German war making in the first half of the twentieth century. From this perspective war is essential to the fulfilment of the national covenant that brings the long-awaited entrance into the Promised Land, a refuge in the imagination that offers what Daniel Silver calls the 'the possibility of escaping tyranny' (in the German case the tyranny of 'Rome').[5] But the faith of the Chosen People in the eventual bestowal of this reward must be continually tried, which is why religious war is perceived to be a series of trials and defeats, a history of covenants made and broken, that can only end in the Final Judgement.[6] In the temporal nationalist consciousness, the nation always remains unfinished.

In the nineteenth century Protestant and Catholic historians refashioned the story of the Thirty Years' War as the great conflict, a singular epic of victimization, which for better or worse laid the foundations of the modern German nation. In constructing this narrative they uncovered, or so they believed, the meaning of centuries of defeat, territorial fragmentation, and political disunity. Interpreting the tribulations of the war as tests of faith (and thus as signs of God's covenant with the German nation), Protestant historians elaborated Germany's martyrdom at the hands of the Counter-Reformation as the sacrifice that revealed the Germans as Europe's nation of destiny. In 1871 they celebrated the fulfilment of the covenant of 1517. This triumph pushed the Catholic story of the nation, against fierce resistance (1871 representing the end of the confederal idea of Germany), to the margins of a new national (read Protestant/Prussian) history.

In 1914 a new imperium briefly beckoned as the glittering prize of Germany's holy war.[7] But the covenant of 1871 was broken in 1918, a defeat that led many Germans, staggered at yet another failure to fulfil the national destiny, to embrace Adolf Hitler's new covenant, the racially pure *Volksgemeinschaft*. After 12 years, the collapse the Thousand Year Reich in 1945 seemed to replay the original catastrophe of 1618–48 by ending, once again, with a disastrous failure to remake the nation in war.

This national narrative of suffering in war, a conflation of martyrdom and eschatological triumph, acquired a unique vocabulary, basically a perception of threat, which contextualizes the Germans' justification of their country's entry into war in 1914. Chancellor Bethmann-Hollweg, in his address to the *Reichstag* at the outbreak of hostilities, spoke passionately about Germany's historical fear of encirclement and occupation when he warned his audience that 'enmity has been awakened against us in the East and the West, and chains have been fashioned for us'.[8] Johannes Kaempf, the president of the *Reichstag*, urged his colleagues to vote for war credits secure in the conviction that 'the war into which we were forced is a defensive war; but at the same time its highest spiritual and material possessions are at stake for Germany. It is a war of life and death, a struggle for existence'.[9] But nowhere in the patriotic and bellicose declarations of 1914 was the experience of the Thirty Years' War referred to more explicitly than in the 'Manifesto of the Ninety-Three German Intellectuals', published under the heading 'To the Civilized World!' (*An die Kulturwelt!*) in the *Kölnische Zeitung* on 4 October 1914. Signed (if not read) by prominent representatives of the German intelligentsia, including Peter Behrens, Lujo Brentano, Adolph von Harnack, Max Lenz, and Gustav von Schmoller, it stridently defended German war: 'It is not true that the combat against our so-called militarism is not a combat against our civilization, as our enemies hypocritically pretend it is. Were it not for German militarism, German civilization would long since have been extirpated. For its protection it arose in a land which for centuries had been plagued by bands of robbers, as no other land has been.'[10] German nationalism not only embraced the nation's suffering in war as unique in scale and injury but also asserted that the stakes of German war had *always* been the life or death of the nation.

Otto Hintze, writing in 1915, declared the rise of the military state of the Hohenzollerns as 'a bitter, perhaps unavoidable necessity' directly traceable to the Thirty Years' War, a comprehensive disaster that had left Germany weak and fragmented at the centre of a continent of covetous and more powerful enemies. Comparing Germany's situation in the Great War to Prussia's during the Seven Years' War, Hintze declared that 'the German Reich, under a Hohenzollern Kaiser, battles for its existence against a world of enemies'.[11] In Hintze's argument we can see the German dread of occupation, partition, and dissolution, coupled with an assertion of the historical connection between national unification and the 'necessity' of the 'monarchical-military factor' in modern German history.[12] From this perspective, it is understandable that World War I raised among many Germans the same fears about German vulnerability (and the catastrophic consequences of defeat) that ran so consistently through the popular memory of the Thirty Years' War. One anodyne to these fears was fealty to the nationalist faith that it was in war and in the military power of the Prussian state that, in the words of Ernst Troeltsch and Friedrich Meinecke, the 'life force' (*Lebenskraft*) and the 'genius' of German civilization manifested themselves.[13] The leader of the wars of unification, Helmuth von Moltke, gloomily predicted towards the end of his life that Germany would eventually have to fight a war lasting 'perhaps seven and perhaps thirty years' that pitted the 'Slav East and the Latin West against the center of Europe'. German military planning before the war confronted this *bête noire* in a preoccupation with the 'encirclement' of Germany, developing strategies that sought quick and decisive battles of annihilation in the opening phases of a general European war.[14] Jay Winter acknowledges this over-mastering German fear in a broader cultural argument, pointing out that World War I was 'the last nineteenth-century war, in that it provoked an outpouring of literature touching on an ancient set of beliefs about revelation, divine justice, and the nature of catastrophe'.[15] Framed within these lines, it is clear that military thinkers in nineteenth-century Germany did not conceptualize the waging of war according to solely (or even primarily) rational principles. A vision of religious war, in which the survival of the nation could only be imagined within the contexts of sacrifice and annihilation, was the background of their approach to war. This conception of war, in turn, legitimated its increasingly radical conduct.

Most Germans, like the other combatant populations, mobilized for war in the summer of 1914 believing that their nation faced a mortal threat to its existence. But in the German case this conventional patriotic spirit was also informed by a nationalist discourse that embraced Herodotus's dictum that 'the gods love to thwart whatever is greater than the rest' when it asserted an eschatological conviction that Germany's wars had always summoned divine judgement.[16] Underlying that conviction (the corollary of Friedrich Schiller's declaration that 'world history is the world tribunal') was the collective memory of the 'time of extermination' (*Vertilgungszeit*), the Thirty Years' War, that had been developed and nurtured in the nineteenth century. In this historical consciousness, *all* war was ultimately religious war because war was, fundamentally, the sacrifice and trial God demanded to fulfil the covenant with his chosen people. This trial, narrated not only in the histories of the Thirty Years' War but also in the Prussian School's histories of the Hohenzollerns, was the history of Germany's rise to nationhood. Properly understood, 'German war' thus could never be entirely separated from a fear of catastrophic defeat and annihilation, the fear of another Thirty Years' War.

This conclusion is reinforced when we recognize that German nationalism evolved out of the Protestant–Catholic divide. Protestant nationalism viewed Luther's defiance of Rome in 1517 as a renewal of the war of independence that began with Hermann's victory over the Roman legions in the Teutoberg Forest in 9 AD. This worldview elevated Germany to the status of defender of European civilization from Latin decadence and Jesuit dictatorship. The Kaiser's failure to re-Catholicize Germany over the course of the Thirty Years' War confirmed Protestant Germany's 'chosen people' narrative and inspired the great awakening of German nationalism in the nineteenth century. In 1893 Georg Winter claimed that the Protestant cause in the war (even if the war had not resulted in a unified state purged of confessional division) was the beginning of a 'genuine German national Protestant idea', a revolutionary moment in German history when 'a great unified spiritual movement (*Bewegung*) went through all Germans in their political disunity'.[17] The nineteenth-century histories of the Thirty Years' War, which focused in part on a sequence of unsuccessful attempts at unification pursued by Tilly, Gustavus Adolphus, and Wallenstein between 1629 and 1634, expose the uncertainties behind nineteenth-century German nationalism and, to a large extent, explain the bellicose self-righteousness so incomprehensible to Germany's enemies during World War I. This attitude, characterized more concretely as Protestant nationalism, a 'liberation theology' emerging out of 1517, also accounts for the incredulity of Germany's politicians and intellectuals that Britain and France (but 'England' especially) would fight on the side of Asiatic Russia, an alliance that philosopher Max Scheler bluntly condemned in 1916 as 'a betrayal of Western culture'.[18] Scheler saw the war as a 'German war', not a 'world war', a war waged not only for Germany's 'existence, independence, and freedom' but also for Europe and, indeed, for all of mankind.[19] In addition, many German intellectuals believed that 1914 marked the renewal of the crusade of 1870–1 against French rationalism, materialism, and republicanism, a conflict that was part of the longer German resistance to French expansionism.[20] German theologians such as Troeltsch and Adolph von Harnack, mystified at how Germany's enemies had lined up against her, could not comprehend Britain's abandonment, in joining France and Russia, of the German Protestant cause of defending European civilization.[21]

This nineteenth-century *Nationalprotestantismus* was a formative dynamic in the 'war theology' that Wilhelm Pressel sees emerging in World War I as an attempt by the church to justify the destruction of the war by preaching a 'nationalist understanding of God'. Bearing a strong resemblance to Scheler's optimism that a new national consciousness would come out of the war, this idea sanctified German war aims as a battle for a divinely ordained temporal order, in this instance the creation of a new *Volksgemeinschaft* morally superior to that of Germany's materialist and spiritually decadent enemies. Like the Thirty Years' War, the Wars of Liberation, and the Wars of Unification before it, the Great War was conceived as a religious war that promised the 'fulfilment' (*Vollendung*), in a millenarian sense, of German history. War delivered the divine judgement on which nation would survive to establish the kingdom of God on earth. It could therefore only be defined as an apocalyptic battle between good and evil.[22] In 1911 the theologian Karl Holl looked back on the Thirty Years' War as the wellspring of 'the [German] conviction of the ethical worth of the state [and] the commitment to the whole which was prepared even for heavy sacrifices. The inner resources of the state were thus greatly increased. Herein lies the reason why the Protestant lands recovered so quickly from the Thirty Years' War . . . why poor Prussia could become a great power'.[23] With a mixture of anxiety and hope Protestant nationalism understood the war of 1914–18 as part of a much longer continuum of struggle that defined the German rise to nationhood. In Wolfgang Mommsen's analysis, the popular notion that Germany was fighting a war in defence of 'German liberty' (*Deutsche Freiheit*) also embraced the latter as embodying a transcendent moral alternative to the 'degenerated individualism' of Western civilization.[24] The German nation, rescuer of Europe, was the last bastion of the Kingdom of God, or, at least, 'true' Western Christendom. Conceiving war as religious war had become, in Michael Jeismann's phrase, an integral part of the 'German self-understanding' of their development as a nation.[25] The German fear of national dissolution, and the hope for unity, could only be reconciled, in Roger Chickering's view, in all or nothing terms: 'Obviously, war, conflict, and battle were fundamental elements in the Pan-German conception of the world.'[26]

Three years of stalemate and mass death had driven home to most Germans the reality that the ambitious designs of 1914 would not be fulfilled. When the stakes are so high, and the cause so righteous, defeat is easily rationalized and assimilated as martyrdom. A catastrophic defeat threatened the covenant of 1871 as the nation seemed once again on the brink of dissolution. The dread of enslavement and annihilation pressed heavily on Germany. Nevertheless another opportunity for rebirth, if not victory, could be glimpsed through the approaching storm. The looming collapse was yet another crisis in a longer struggle that had begun in 1618 (if not 1517), one more trial and judgement of the German nation by God. Karl Holl announced to an assembly of Lutheran missionaries and church leaders in Vilna in 1916 that the world war must be understood as a religious war sent by God to test the 'moral worth' (*sittliche Wert*) of the German people. Germans could summon the spiritual power to prevail, Holl believed, if they recalled how the nation had twice before fought a 'battle for existence' (*Daseinskampf*): the Thirty Years' War and the Wars of Liberation. For Holl the historical and spiritual parallels between the Thirty Years' War and the world war were obvious – both were wars of 'to be or not to be' (*sein oder nicht sein*) for German Protestantism.[27] Echoing Holl, a collection of contemporary accounts of the Thirty Years' War published in 1917 opened with the foreword to the 1683 edition of Grimmelshausen's *Simplicissimus*. Titled 'To the Loyal German Reader', it reminded Germans of the destruction of the Thirty Years' War: burned villages, destroyed churches, raped women, and German blood that had 'flowed like water' – all the sacrifices demanded in the defence of German liberty from the 'tyranny of Byzantium'.[28]

Mirroring all the broken covenants in German history, the defeat of November 1918 had followed stunning victories – in this instance the collapse of Russia in 1917 that promised the fulfilment of the *Drang nach Osten*. 'An imperium of grandiose dimensions', as Fritz Fischer described it, seemed after terrible trials to be suddenly and miraculously at hand. At a stroke, a 'contemptible' (*ordinär*) negotiated peace need no longer be contemplated, now that the entire weight of the German army could be brought to bear on the Western Front.[29] Given the state of German morale, manpower, and resources, the offensives Ludendorff launched in the spring and summer of 1918 have usually been characterized as a 'last throw of the dice' – a desperate bid to win the war before American intervention erased all hope of a conclusion on German terms. However, another scenario strongly suggests itself. The 1918 offensives can also be seen as the defining act of German war as religious war: the supreme moment in the struggle against Amalek in which only one nation, the chosen people, would survive. Post-war reflections on the collapse in the late summer and autumn of 1918 bear this out. Treatises such as Günther Frantz's *Die Vernichtungsschlacht* (*Battle of Annihilation*; 1928) and Waldemar Erfurth's *Der Vernichtungssieg* (*Annihilative Victory*; 1938) eagerly embraced the notion of national rebirth in defeat and destruction.[30] Erich Ludendorff, in his 1935 book *Total War*, defined the latter as 'the highest expression of the [nation's] racial will to live'.[31] Michael Geyer's profoundly apt term 'catastrophic nationalism' succinctly encompasses the dread that darkened the counsels of the German High Command towards the end of 1918, an apprehension that went beyond fear of defeat on the battlefield. The military and civil collapse in 1918 threatened the dissolution of Germany in Bolshevik-style revolution. Faced with this grim prospect, some members of the leadership sought to snatch a moral victory from the jaws of defeat and called for a popular uprising (*Volksaufstand*) in the tradition of 1813 – a 'war unto destruction' (*Endkampf*) that promised national rebirth.[32] But reality soon overcame Ludendorff's Wagnerian visions of *Götterdämmerung*. Ultimately, the generals meekly handed power to the civilians and, following their Kaiser's example, retired from the field, only to return after the peace to inveigh against the fall of civilization they saw looming in the epochal 'world civil war' unleashed by Lenin in 1917.[33] Above all, they preached the necessity of German moral regeneration in conditions 'reminiscent of the period after the Thirty Years' War'.[34] It is clear that the broken covenant of 1918 generated a new level of radicalized nationalism in which violent fantasies of national redemption found fertile ground.

The narrative of German history was inconceivable without the beginning story of defeat – indeed, the narrative in all its versions, Protestant and Catholic, was a *response* to defeat. Set down in the histories of the Thirty Years' War, the defining national epic, this narrative was written in what Stefan-Ludwig Hoffman, in his study of the commemorations of the Leipzig 'Battle of the Nations' in 1813, calls 'the pietistic language of liberation': a nationalist rhetoric unable to free itself from the morbid conviction that German history was an unending cycle of subjection, sacrificial death, and rebirth.[35] Annette Maas points out that rarely has the story of defeat been reconciled with the imperatives of nationalism without recourse to Christian imagery promising redemption in the covenant, the establishment of the Kingdom of God on earth.[36] This confluence of religious millenarian hope, sacral nationalism, and the trauma of mass death continued to exert a powerful influence on the conception of the German nation after the end of World War I. In his classic study of mass death and German memory after the world wars, George Mosse concluded that after 1918 invocations of the 'holy nation' consistently drew on the imagery and vocabulary of religious war and the purifying pagan sacrifice of Siegfried. Here we see the German national narrative shaped by the classical introspection of thanatopsis, or a liturgical meditation on death, revealed in Hermann Oncken's apprehension of 'the dead of 1914–1918 who like a vast army of ghosts float between the old and the new Germany, between our past and our future'.[37] In the traumatic memory of mass death radical nationalist thought, with all of its extravagant promises of the rebirth of a 'new' Germany, found its essential medium. It promised not only the recovery of what was lost in the defeat but also the resurrection from what Pan-German leader Heinrich Class identified as the 'moral breakdown' (*sittliche Zerfall*) and 'cultural degeneration' (*kulturellen Rückgang*) that had brought on defeat, not just the resurgence of national pride and power, but a *völkisch-national* reawakening manifested in territorial expansion and a fundamental alteration of the political system towards dictatorship.[38] Radical German nationalism discovered in defeat a revolutionary moment to remake the nation.

After 1918, nationalist thinkers like Class revisited the moral themes of regeneration and renewal that had been fundamental to the nineteenth-century histories of the Thirty Years' War. In 1920 Ricarda Huch offered her readers a fictionalized 'character study' of Wallenstein, the imperial generalissimo immortalized by Friedrich Schiller, which described the general's struggle to resurrect a 'degenerate and decadent' Reich through war and violent revolution.[39] In post-1918 Germany, where extremist advocacy of regeneration through violence sought political and ideological traction in defeat and revolution, Huch's reappraisal of the career of Wallenstein was a sign of the times. Alfred Döblin's *Wallenstein* (also published in 1920), his only work not banned by the Nazis, found favour in the Nazi worldview, Harro Müller believes, because it advanced a view of history as 'war history [which] is the history of terror with strongly marked cyclical moments'.[40] In 1936 Walther Tritsch declared that 'the theme of Wallenstein is in the air', comparing Wallenstein's use of war to unify the nation to Adolf Hitler's vision of a revitalized Germany.[41] Carl Rummel, writing in 1941, interpreted the Thirty Years' War as a providential moment when Wallenstein, 'a bridge between past and future', attempted to 'remold the Reich into a new form'. Rummel believed that Adolf Hitler had been anointed to complete Wallenstein's unfinished project.[42]

The fundamental idea behind the Nazi 'revolution' was its vision of national rebirth driven by racial renewal. World War I and the ten million dead it left in its wake revived the fears of racial decline that had informed pre-war Europe's social Darwinist worldview. After 1918 pro-natal and eugenics movements and explicit racial definitions of national health and vitality found many adherents, not all of whom were on the radical fringe.[43] References to the enormous population losses (and forced migrations) of the Thirty Years' War promoted a vision of German recovery after World War I premised on the necessity of racial regeneration. A noted historian of the Holy Roman Empire, Karl Brandi, asserted in his 1927 history of the Thirty Years' War that Germany's suffering in that conflict strengthened the 'national soul' (*Volksseele*), leading to a profound and enduring 'rediscovery of the national'.[44] Nineteenth-century ideas of Nordicism and racial renewal, which shaped the Nazi concept of the racially pure national community (*Volksgemeinschaft*), acquired new persuasiveness in the wake of defeat. New interpretations of the German campaign of the Swedish King Gustavus Adolphus between 1630 and 1632 reflected these notions. Johannes Paul, the Protestant hero's pre-eminent twentieth-century biographer, pointed out that, in light of the damage done to Germany by Napoleon, perhaps it would have been better if a 'racially kindred hero' (*Stammverwandter Held*) had reorganized the Holy Roman Empire as a 'great Swedish-German Reich'.[45] Ernst Kohlmeyer, writing in 1940, seconded Paul when he claimed that a 'strongly bound Germanic Reich' (*germanischen, festumgrenzten Reich*) led by a 'racially kindred blood relative' (*stammesnäher, blutsverwandter*) would certainly have been better for Germany than the tyranny of 'the international, half-Spanish imperium with its Croats and Magyars'.[46] In 1932, on the tercentenary commemoration of the king's death at Lützen, a young Gerhard Ritter wrote in his student newspaper *Wingolfsblätter* that such a state, a Nordic bulwark as it were, would have shielded German-Protestant culture from 'Muscovite barbarism'.[47] A year later Richard Schmidt compared Gustavus Adolphus's unsuccessful attempt to create a Swedish-German union to defend Protestant northern Europe with Alexander the Great's failure to unite the 'Hellenistic-Asiatic world' against Persia.[48] In 1943, as the Third Reich faced almost certain defeat, Günther Franz exhorted war-weary Germans to remember that the Thirty Years' War, for all its destruction, had created a more vital 'German racial stock' (*deutschen Volkstämme*).[49]

Even before the war, an interpretation of the legacy of the Thirty Years' War that supported Adolf Hitler's racial worldview received the imprimatur of the Nazi Party. In December 1937 Wilhelm Frick, the Nazi Minister of the Interior, gave a speech to the German-Swedish Society wherein he urged the construction of a new national monument to Gustavus Adolphus on the site of the existing memorial chapel (built in 1907) at Lützen. Racializing the popular homage to the Protestant hero, Frick declared that the 'true Germanic hero king' had been fighting, even if unconsciously, for the 'Nordic-Germanic racial idea' – a struggle on behalf of 'the entire Nordic-Germanic worldview' for which Adolf Hitler crusades today.[50] Frick's speech was used as a preface to a 1939 book by Hans Chilian that posed the question, 'What present day significance does the king have for the racial consciousness of Sweden and Germany?'. According to Chilian, the answer lay in the revelation that Gustavus Adolphus's spirit 'walks in the ranks of all German freedom fighters, in the army of Frederick the Great, the Wars of Liberation, the Wars of Unification, and the World War . . . inspiring the racial comrades (*Volksgenossen*) of the Third Reich with new energy [in the battle] against Bolshevism and imperialism'. Chilian had no doubt that there was 'a direct line from Gustavus Adolphus to Adolf Hitler'.[51]

Adolf Hitler himself made frequent references to the 'fifteen million dead' of the Thirty Years' War as evidence of the struggle in which the German *Volk* proved its worth. Hitler had originally believed, as he wrote in *Mein Kampf*, that the war had caused a prolonged decline of the German race, when 'poisonings of the blood' had led to 'a decomposition of . . . our soul'.[52] But as he rose to power, Hitler pointed more and more to the demographic recovery from the losses of the Thirty Years' War (referring always to the figure of 15 million dead) as proof of German racial vitality and superiority.[53] After the outbreak of war in 1939 these references were a frequent theme in his wartime table talk.[54] On 4 July 1944 in Obersalzberg, in one of his last appearances before a non-military audience, Hitler rallied himself and 200 industrialists while, 800 kilometres to the east, the Soviet Army was destroying Army Group Centre. Looking back on the 'numberless wars', including the Thirty Years' War, which the German people had survived, 'even as they would survive this one', Hitler reminded the magnates that 'it is always better that the birth certificate of a new Reich be written in blood, with blood and in crisis'.[55]

A commonplace in the morbid *fin de siècle* cultural environment that formed Hitler's worldview was the social Darwinist notion that war was essential to the survival and improvement of the race. Richard Evans believes, however, that social Darwinism in Germany was only one idea within a more complex national self-conception that understood historical and social progress in terms of conflict and struggle.[56] The German conception and waging of war were, as Manichean and annihilative constructs, the extreme manifestations of this idea, which originated in what was essentially the story of a great defeat: the Thirty Years' War. The year 1918 was seen as a parallel catastrophe. The persuasiveness of this point of reference helps explain the genocidal vision of a German-ruled Nordic empire created by the enslavement and mass murder of Jews and Slavs in a war waged in defence of true Christian culture.[57] It is at this intersection of war, defeat, and revolution that, to use Mark Mazower's terminology, the 'dialectic of Nazi racial warfare' emerges. The most compelling aspect of Mazower's analysis is not the conventional conclusion that the Thousand-Year Reich could only be founded in war, but his assertion that the Nazi drive for *Lebensraum* arose out of the post-1918 German fear of 'folk death', which I believe had been elevated as a central feature of the German national consciousness by the nineteenth century's 're-discovery' of the Thirty Years' War.[58] Protestant histories of the war clearly conceptualized the German struggle for nationhood from the perspective of religious war. A tragic, though admittedly an unintended, consequence was that these histories effectively legitimated radical war making which, under Adolf Hitler's direction, encompassed genocide.[59]

The war that Hitler launched in 1939 was conceived as a religious war, a crusade, in which millions of Jews, Slavs, and other racial enemies of the Reich were to be extinguished from the memory of men. If we are to fully understand the destructive energies that propelled the Third Reich towards mass murder and *Götterdämmerung*, we must look closely at how the German national consciousness assimilated the course of history as the story of covenants and sacrifice. In the nineteenth century German historians took the story of the Thirty Years' War and remade it into a national narrative preoccupied with the fear of defeat and annihilation. This was history writing as thanatopsis, in which Norbert Elias sees the expression of the German quest for 'fulfillment in destruction'.[60] This quest is, ultimately, the story of German war in the twentieth century.

Notes

1. Carl von Clausewitz, *On War*, ed. and trans., Michael Howard and Peter Paret (Princeton, NJ: Princeton University Press, 1984), p. 605.
2. Ibid., pp. 80–1, 593, 609.
3. See Kevin Cramer, *The Thirty Years' War and German Memory in the Nineteenth Century* (Lincoln, NE: The University of Nebraska Press, 2007).

4. See Kevin Cramer, 'Religious Conflict as History: The Nation as the One True Church' in Michael Geyer and Hartmut Lehmann (eds), *Religion und Nation: Beiträge zu einer unbewältigte Geschichte* (Göttingen: Wallstein Verlag, 2004), pp. 23–38.
5. Daniel J. Silver, *A History of Judaism. Volume I: From Abraham to Maimonides* (New York: Basic Books, 1974), p. 40.
6. See Kenneth Burke, *The Rhetoric of Religion: Studies in Logology* (Berkeley, Los Angeles, CA, and London: University of California Press, 1970), pp. 174–81. Burke opens this discussion with a quote from Coleridge: 'A fall of some sort or another . . . is the fundamental postulate of the moral history of man. Without this hypothesis, man is unintelligible; with it, every phenomenon is explicable.'
7. See Kevin Cramer, 'A World of Enemies: New Perspectives on German Military Culture and the Origins of the First World War', *Central European History*, vol. 39, no. 2 (June 2006), 270–98.
8. Theodor von Bethmann-Hollweg, 'Speech of the Imperial Chancellor, Dr. von Bethmann-Hollweg, in the Reichstag August 4, 1914' in Ralph Haswell Lutz (ed.), David G. Rempel and Gertrude Rendtorff (trans.), *Documents of the German Revolution: The Fall of the German Empire, 1914–1918* (Stanford, CA: Stanford University Press, 1932), pp. 1, 10.
9. Johannes Kaempf, 'Speech of the President of the Reichstag, Dr. Kaempf, August 4, 1914' in Lutz, *Documents of the German Revolution*, p. 14.
10. Peter Behrens et al., 'The Manifesto of the German University Professors and Men of Science' in Lutz (ed.), *Documents of the German Revolution*, p. 75. On the manifesto, see Ernest A. Menze, 'War Aims and the Liberal Conscience: Lujo Brentano and Annexationism during the World War I', *Central European History* 17, nos 2/3 (June/September 1984), 140–58.
11. Otto Hintze, *Die Hohenzollern und ihr Werk. Fünfhundert Jahre väterlandischer Geschichte* (Berlin: P. Parey, 1915), pp. vii, 187, 685.
12. Otto Hintze, 'Deutschland und das Weltstaatensystem' in Otto Hintze, Friedrich Meinecke, Hermann Oncken and Hermann Schumacher (eds) *Deutschland und der Weltkrieg*, 2nd and rev. edn (Leipzig and Berlin: B. G. Teubner, 1916), pp. 11–12.
13. Ernst Troeltsch, 'Der Geist der deutschen Kultur' and Friedrich Meinecke, 'Kultur, Machtpolitik, und Militarismus' in Hintze, Friedrich Meinecke, Hermann Oncken and Hermann Schumacher (eds), *Deutschland und der Weltkrieg*, 2nd and rev. edn (Leipzig and Berlin: B. G. Teubner, 1916), pp. 71 and 757.
14. Gunther E. Rothenburg, 'Moltke, Schlieffen, and the Doctrine of Strategic Envelopment' in Peter Paret (ed.), *The Makers of Modern Strategy from Machiavelli to the Nuclear Age* (Princeton, NJ: Princeton University Press, 1986), pp. 305–6. See also Terence Zuber, *Inventing the Schlieffen Plan: German War Planning, 1871–1914* (Oxford: Oxford University Press, 2002).
15. Jay Winter, *Sites of Memory, Sites of Mourning: The Great War in European Cultural History* (Cambridge: Cambridge University Press, 1998), pp. 179–8.
16. Herodotus, *The History*, trans. David Greene (Chicago and London: The University of Chicago Press, 1987), p. 472.
17. Georg Winter, *Geschichte des Dreißigjährigen Krieges* (Berlin: G. Grote, 1893), pp. 4–6.

18. Max Scheler, 'Der Genius des Krieges und der Deutschen Krieg [1916]' in Manfred S. Frings (ed.), *Politisch-Pädagogische Schriften* (Bern and Munich: Francke, 1982), p. 141.
19. Max Scheler, *Krieg und Aufbau* (Leipzig: Verlag der Weissen Bücher, 1916), p. 14; Scheler, 'Genius des Krieges', pp. 137–9.
20. Martin Greschat, 'Krieg und Kriegsbereitschaft im deutschen Protestantismus' in Jost Düffler and Karl Holl (eds), *Bereit zum Krieg: Kriegsmentalität in Wilhelmischen Deutschland, 1890–1914* (Göttingen: Vandenhoeck & Ruprecht, 1986), p. 35.
21. Hew Strachan, *The First World War*. Vol. 1, *To Arms* (Oxford: Oxford University Press, 2001), p. 1116.
22. Wilhelm Pressel, *Die Kriegspredigt 1914–1918 in der evangelischen Kirche* (Göttingen: Vandenhoeck & Ruprecht, 1967), pp. 21–6, 51, 81–3, 140–53.
23. Karl Holl, *The Cultural Significance of the Reformation* [1911], 2nd rev. edn (1948), trans. Karl and Barbara Hertz and John H. Lichtblau (New York: Meridian Books, 1959), p. 57.
24. Wolfgang J. Mommsen, *Bürgerliche Kultur und politische Ordnung: Künstler, Schriftsteller und Intellektuelle in der deutsche Geschichte 1830–1933* (Frankfurt: Fischer Taschenbuch Verlag, 2000), pp. 178–81.
25. Michael Jeismann, *Das Vaterland der Feinde: Studien zum nationalen Feindbegriff und Selbstverständnis in Deutschland und Frankreich, 1792–1918* (Stuttgart: Klett-Cotta, 1992), pp. 301–2.
26. Roger Chickering, 'Die Alldeutschen erwarten den Krieg' in Dülffer and Holl (eds), *Bereit zum Krieg: Kriegsmentalität in Wilhelmischen Deutschland, 1890–1914* (Göttingen: Vandenhoeck & Ruprecht, 1986), pp. 23–5.
27. Karl Holl, *Die Bedeutung der großen Kriege für das religiöse und kirchliche Leben innerhalb des deutschen Protestantismus* (Tübingen: J.C.B. Mohr, 1917), pp. 4–5.
28. Johann Jonathan Felßecker, 'Teutsch-treugesinnter Leser' in Hans Karl Schulz (ed.) *Der Dreißigjährige Krieg*, vol. 1, *Bis zum Tode Gustav Adolfs*, Hauptquellen zur neueren Geschichte, (ed.) Erich Brandenburg (Leipzig and Berlin: B.G. Teubner, 1917), pp. 1–2.
29. Fritz Fischer, *Germany's Aims in the World War I*, trans. James Joll (New York: W.W. Norton, 1967), pp. 607–13.
30. Jehuda Lothar Wallach, *The Dogma of the Battle of Annihilation: The Theories of Clausewitz and Schlieffen and Their Impact on the German Conduct of Two World Wars* (Westport, CT, and London: Greenwood Press, 1986), p. 218.
31. Erich Ludendorff, *Der Totale Krieg* (Munich: Ludendorffs Verlag, 1935), quoted in Wallach, *The Dogma of the Battle of Annihilation*, p. 240.
32. Michael Geyer, 'Insurrectionary Warfare: The German Debate about a *Levée en Masse* in October 1918', *Journal of Modern History* 73, no. 3 (September 2001), 467, 473–6, 509–14.
33. Dan Diner and Bill Templar, 'European Counter-Images: Problems of Periodization and Historical Memory', *New German Critique*, no. 10 (Spring 1985), 165–6.
34. Richard Bessel, *Germany after the First World War* (Oxford: Oxford University Press, 1995), pp. 220, 243.
35. Stefan-Ludwig Hoffmann, 'Mythos und Geschichte: Leipziger Gedenkfeieren der Völkerschlacht in 19. und 20. Jahrhundert' in Etienne François, Hannes Siegrist, and Jakob Vogel (eds) *Nation und Emotion: Deutschland und Frankreich*

in Vergleich 19. und 20. Jahrhundert (Göttingen: Vandenhoeck & Ruprecht, 1995), p. 114.
36. Annette Maas, 'Der Kult der toten Krieger. Frankreich und Deutschland nach 1870/71' in *Nation und Emotion*, p. 227.
37. George L. Mosse, *Fallen Soldiers: Reshaping the Memory of the World Wars* (New York and Oxford: Oxford University Press, 1990), pp. 34–5, 74–9. On the Siegfried theme, see also Robert W. Gutman, *Richard Wagner: The Man, His Mind, and His Music* (San Diego, CA, New York, and London: Harcourt Brace Jovanovich, 1990), pp. 119–21, 243–4.
38. Ranier Hering, '"Des Deutschen Volkes Wiedergeburt": Völkischer Nationalismus und politische Erneuerungspläne', *Zeitschrift für Geschichtswissenschaft* 42, no. 12 (1994), 1039, 1082–3.
39. Ricarda Huch, *Wallenstein: Eine Charakterstudie* (Leipzig: Insel-Verlag, 1920), pp. 5–6, 12.
40. Harro Müller, 'War and Novel: Alfred Döblin's "Wallenstein" and "November 1918"' in Bernd Huppauf (ed.), *War, Violence, and the Modern Condition* (Berlin and New York: Walter de Gruyter, 1997), pp. 293–7.
41. Walther Tritsch, *Wallenstein: Herr des Schicksal—Knechte der Sterne* (Leipzig: J. Kittl, 1936), pp. 11, 22–5.
42. Carl Johannes Rummel, *Kaiser, Gott, und Reich* (Berlin: Vier Falken Verlag, 1941), pp. 5–6, 820.
43. On nineteenth-century antecedents, see Peter Gay, *The Cultivation of Hatred* (New York: W.W. Norton, 1993) and George L. Mosse, *Nationalism and Sexuality: Respectability and Abnormal Sexuality in Modern Europe* (Madison, WI: University of Wisconsin Press, 1985). Also see William H. Schneider, *Quality and Quantity: The Quest for Biological Regeneration in Twentieth-Century France* (Cambridge and New York: Cambridge University Press, 1990) and Claudia Koonz, *Mothers in the Fatherland: Women, the Family, and Nazi Politics* (New York: St. Martin's Press, 1987).
44. Karl Brandi, *Die Deutsche Reformation. Gegenreformation und Religionskriege. 2. Halbband* (Leipzig: Quelle & Meyer, 1927), pp. 306–9.
45. Johannes Paul, *Gustav Adolf: Erster Band: Schwedens Aufsteig zur Grossmachtsstellung* (Leipzig: Quelle & Meyer, 1927), p. 6. In 1930 Paul referred to this as a 'North German-Scandinavian Unity Front', see Johannes Paul, 'Gustav Adolf in der deutschen Geschichtsschreibung', *Historische Viertaljahrschrift* 25, no. 3 (September 1930), 429.
46. Ernst Kohlmeyer, *Gustav Adolf und Deutschland* (Bonn: Bonner Universitäts-Buchdruckerei, 1940), pp. 11–13.
47. Gerhard Ritter, 'Gustav Adolf, Deutschland und das nordische Luthertum [1932]', chap. in *Die Weltwirkung der Reformation*, 2nd edn (Munich: R. Oldenbourg, 1959), pp. 137–8, 144–5.
48. Richard Schmidt, 'Gustav Adolf. Die Bedeutung seiner Erscheinung für die europäische Politik und für den deutschen Volksgeist', *Zeitschrift für Politik* 22, no. 11 (February 1933), 719.
49. Günther Franz, *Der Dreißigjährige Krieg und das deutsche Volk: Untersuchungen zur Bevölkerungs-und Agrargeschichte*, 2nd rev. edn (Jena: G. Fischer, 1943), p. 100.
50. Wilhelm Frick, foreword to Hans Chilian, *'Geknechtete befreite er!' Gustav Adolfs nordischer Freiheitskampf im Lichte unserer Zeit* (Leipzig: G. Kummer, 1939), pp. vii–xv.

51. Hans Chilian, *'Geknechtete befreite er!'*, pp. 211, 213, 220–1.
52. Adolf Hitler, *Mein Kampf*, trans., Ralph Mannheim (Boston: Houghton Mifflin, 1971), p. 396.
53. Adolf Hitler, 'Rede vor dem Industrieklub in Düsseldorf [1932]' in Max Domarus (ed.), *Hitler. Reden und Proklamationen, 1932–1945. Band I: Triumph: Erster Halbband, 1932–1934*, (Munich: Süddeutscher Verlag, 1965), p. 71; and 'Regensburger Rede [1937]'; and 'Aufbau und Organisation der Volksführung [1937]' in *Band I: Triumph: Zweiter Halbband, 1935–1938*, pp. 699–701 and 761.
54. Adolf Hitler, 'Tischgespräche, Juli 1941-August 1942' in Domarus, *Band II: Untergang: Zweiter Halbband, 1941–1945*, p. 1744.
55. Adolf Hitler, 'Tagung von Wirtschaftführern [1944]' in Domarus, *Untergang: Zweiter Halbband*, p. 2115.
56. Richard J. Evans, 'In Search of German Social Darwinism', chap. in *Rereading German History: From Unification to Reunification, 1800–1996* (London and New York: Routledge, 1997), p. 137.
57. For a fascinating look into how this vision was articulated in German science fiction between 1920 and 1945, see Jost Hermand, *Old Dreams of a New Reich* (Bloomington and Indianapolis, IN: Indiana University Press, 1991), pp. 246–63.
58. Mark Mazower, *Dark Continent: Europe's Twentieth Century* (New York: Alfred A. Knopf, 1999), pp. 79–80, 157–60.
59. On the influence of Imperial Germany's colonial campaigns on the Third Reich's genocidal war making, see Isabel Hull's *Absolute Destruction: Military Culture and the Practices of War in Imperial Germany* (Ithaca and London: Cornell University Press, 2005).
60. Norbert Elias, *The Germans: Power Struggles and the Development of Habitus in the Nineteenth and Twentieth Centuries*, Michael Schröter (ed.), trans., Eric Dunning and Stephen Mennell (New York: Columbia University Press, 1996), pp. 208, 320–1.

7

The Stories of Defeated Aggressors: International History, National Identity and Collective Memory after 1945

Patrick Finney

The old adage that 'history is written by the victors' conveys at best a partial truth. Except in limit case of absolute annihilation, defeated states are seldom precluded from historicizing their predicament through formal and informal discourses of memory, and indeed the dislocating experience of catastrophic reversal may well render the narrative balm of history all the more imperative. The explanations propagated will inevitably be constrained by a range of political, practical and psychological factors, and of course not all interpretations have equal suasive power. Yet it does not automatically follow either that the readings of the defeated will be bashful and neutered or that they will fail to gain a broader purchase. Modern international history provides the perfect illustration of this in the post-1919 debate over responsibility for the outbreak of the First World War. The Allied and Associated Powers quite literally wrote history in Article 231 of the Treaty of Versailles, the so-called war guilt clause, in which they asserted that the conflict had been 'imposed upon them by the aggression of Germany and her allies'.[1] Although the Germans were forced to acquiesce in this when signing the treaty, they immediately launched a revisionist campaign to undo both the historical claim and the punitive sanctions which depended upon it. By the 1930s their sedulous propaganda had borne fruit, and counter-arguments contending that Europe had simply drifted into war, at the mercy of impersonal systemic forces or fate, were established as orthodox in historiography and in broader collective memory. This in turn played a part in eroding the legitimacy of Versailles and creating a climate conducive to successful Nazi expansionism.[2]

There have been numerous scholarly attempts to analyse and compare 'the historiography of victors and of the vanquished'.[3] These suggest that the interplay between the two is often far from straightforward and that they should not necessarily be seen as dichotomous opposites, perpetually in tension. For example, the Nuremberg and Tokyo war crimes trials were certainly resented by many Germans and Japanese as a flagrant imposition of victors' justice, but at the same time the conspiracy paradigms adopted by Allied prosecutors located responsibility for wartime aggression squarely with small leadership elites, thus by implication exonerating the remainder of the population and closing the book on retribution. So in a broader sense the Allied approach did not discomfit dominant German and Japanese modes of viewing the past, wherein the self-pitying sentiment that they too had primarily been victims, suffering in consequence of the machinations of omnipotent criminal leaders, was firmly entrenched.[4] That said, we should also note that memory among the defeated (or, indeed, the triumphant) is never unitary or uncontested. Where once scholars of war memory might have focused solely on elite public representations and would-be dominant narratives as the basis for judgements about the memories of this or that nation, the burgeoning literature now encompasses myriad cultural forms and the experiences of many sub-national collectives demarcated by class, gender, religion, region, occupation, ethnicity and so on. Knowledge of how these diverse groups consumed, negotiated or rejected official discourses underlines the multiplicity of memory and its status as a domain of ideological struggle, rendering easy generalizations in any national case problematic.

This chapter explores aspects of memory in each of the three main defeated Axis states – the Federal Republic of Germany (FRG), Italy and Japan – after the Second World War, and the contribution of one particular vector of memory, historians of international relations (generally known as diplomatic or international historians). These countries are interesting because in the aftermath of the war, negotiating the recent past meant reckoning not only with lacerating defeat and abject failure but also a potent legacy of guilt. Each state had pursued aggressive expansionist policies that were bound ultimately to call forth resistance, and thus bore signal responsibility for unleashing the conflict. Moreover, albeit to different extents, each carried a burden of culpability not only for the suffering consequently experienced by its own citizens but also for inflicting unprecedented criminal iniquities upon diverse racial and political others. (All this said, however, whether and in what respects just cause for guilt actually existed was to be very much in question in the post-war mnemonic manoeuvring.) In these circumstances, a good many of the consolatory tropes identified by Wolfgang Schivelbusch as typically occurring in cultures of defeat – such as transforming defeat into moral victory or uniting the nation around a programme of *revanche* – were not readily available for deployment.[5]

The repudiation of the actions and values of now defunct regimes provided the general context in which memory settlements were elaborated in each case. The fact that each country experienced occupation by Allied forces, which also oversaw the initial stages of post-war reconstruction, was also significant. The normative expectations of the Allied powers and of wider international public opinion set definite limits on the kinds of narratives that could be put into circulation, mandating a considerable degree of apologetic repentance. However, the defeated needed nonetheless to salvage something from the years of dictatorship and war – to find some way of framing them – that would provide a foundation for the reconstruction of a positive sense of national identity. This imperative to endow traumatized communities with some cohesion and purpose meant that dominant collective memories were thoroughly equivocal, combining gestures of remorse and condemnation with large measures of evasion and self-exculpation. Moreover, as the nascent Cold War became the dominant Allied preoccupation, these former defeated enemies acquired an additional identity as contemporary partners in the struggle against Soviet communism, and this only increased the latitude for the promulgation and indulgence of interpretations that elided recent crimes.

Historians are fond of positing a fundamental distinction between the sober, documented, disciplined narratives that they produce and the interested myths that comprise broader collective memory and contribute to national identity formation. Yet in these cases they are difficult to disentangle. Pieter Lagrou has noted that often after 1945 'scholarly histories were no more than erudite derivatives of political memories', and even once archives opened and mature historiographies developed, history, collective memory and discourses of identity remained inextricably imbricated.[6] From this critical perspective, the contribution of international historians in these three cases is particularly noteworthy. Socially, professionally and politically conservative as a caste, and steeped in the methodological and philosophical traditions of historicism, international historians in these countries had long tended to identify with the nation state, empathetically reconstructing the ideas and actions of statesmen and thereby rationalizing and justifying national policy. By dint of professional specialism, it was this group above all others that confronted the problematic task of providing explanations for pre-war aggression that responded effectively to all the conflicting imperatives in play. Ultimately, their work was to contribute if not a blatantly exculpatory then at least a decidedly conservative strand to the complex yarn of national war memory.[7]

Recently, verdicts on how fully and frankly West Germans came to terms with the Nazi past after 1945 have become more disparaging. Once the FRG was regarded as a model among former Axis powers for its establishment of a healthy democratic culture in which the war and its crimes were 'not only remembered' but 'actively worked on, labored, rehearsed'.[8] The recent memory literature, however, penned in a post-Cold War climate where there is unprecedented candour about Nazi crimes and German complicity, rather stresses inadequacies, omissions and evasions, especially in the first post-war decades. Then, a series of tropes were deployed that appeared to accept responsibility for Nazi crimes but actually elided the issues of complicity and the longer-term roots of aggression within German history. The passive voice was extraordinarily useful here: witness the obfuscation of agency in the archetypal admission that 'unspeakable crimes have been committed in the name of the German people'.[9] The dominant view was that only a handful of perpetrators had really been culpable for Nazi crimes and that, if anything, the generality of Germans had been victims of Nazism too. 'We've been led by criminals and gamblers', lamented a celebrated Berlin diarist in the last desperate days of the war, 'and we've let them lead us like sheep to the slaughter.'[10] Indeed, apology often seemed to be construed as a necessary but tiresome precursor to facilitate an overriding focus on the suffering of Germans, including expellees, refugees and prisoners of war still held in the Soviet Union. Crimes perpetrated by Germans had an extraordinarily low profile, and the racial nature of the whole Nazi project and its defining enormity – the Holocaust – were never brought into focus.[11] A final tactic for displacing responsibility was to present Nazism as an aberration in German history: rather than the product of any 'pressing necessity' peculiar to Germany, it was caused by the pathologies of modernity – 'the optimistic illusions of the Age of Enlightenment and the French Revolution' – and had 'analogies and precedents in the authoritarian systems of neighboring countries'.[12]

The prevalence of these attitudes was part and parcel of a conservative reconstitution of German identity in the post-war FRG. The electoral victory of Konrad Adenauer's Christian Democrats in 1949 signified the triumph of more limited conceptions of German guilt and the prioritizing of reconstruction over retribution. 'The price for post-war integration of those [many] Germans compromised by their beliefs and actions in the Third Reich was silence about the crimes of that period', Jeffrey Herf has argued: 'justice delayed – hence denied – and weakened memory' were required to build this new democracy.[13] The Western Allies connived with alacrity in the curtailing of punishment and re-education, viewing it as the *quid pro quo* for the incorporation of the FRG into the Cold War anti-communist alliance. German historians were also primed to participate in this ideological project. Most had had little difficulty reaching an accommodation with the Third Reich, since they 'had long cultivated their own brand of cultural and political nationalism which proved to be largely compatible with the biological nationalism of the Nazis'. Hence if the profession was not 'Nazified', this was largely because it was unnecessary; without uncritically embracing Nazi ideology, historians nonetheless did collectively legitimate the regime. Moreover, there was considerable continuity through into the post-war period with little purging of personnel or reassessment of methodology or politics. Behind 'a rhetorical smokescreen of lament' the 'same old national apologias, somewhat turned down in volume, could and did continue'.[14]

Historians writing on international relations certainly played a full part in the evasions and suppressions of the Adenauer era. Gerhard Ritter, first chair of the FRG historical association, argued that the core negative elements in Nazism were imports from specific foreign locations – for example, racism from Austria and Machiavellianism from Italy – or simply manifestations of broader European phenomena. He also bluntly defended efforts to throw off the shackles of Versailles and assert the German right to self-determination (as recognized at Munich): it was only when a demonic force turned German policy onto a new and deeply dangerous track with the occupation of rump Czechoslovakia in March 1939 that things began to go awry. This formulation was eminently compatible with an emphasis on German victimhood. Golo Mann – actually among the more democratically minded in the profession – presented Hitler as rising almost from nowhere to beguile a peace-loving people who then suffered both from Nazi tyranny and Allied atrocities. For Mann, blame for the outbreak of the war was located squarely with Hitler and his cabal rather than with the German people who had struggled to resist him; moreover, in explaining its origins space should also be found for British appeasement, American neutrality and Stalin's rapacious pact making in 1939.[15] Such efforts to displace responsibility echoed the self-justifications of Nazi policymakers and shaded into another mode of exoneration stressing the impersonal machinations of geopolitics. For Ludwig Dehio, the war was the product of 'the daemonic nature of power' that impelled states to seek hegemony, and of the tragic contingency that allotted Germany the physical location, means and opportunity to play this doomed part. Once the state was seized by the 'Satanic genius' of Hitler, the die was cast.[16] This argument relativized Nazi actions in the context of an amoral international system and almost eradicated human (and German) agency by focus on structural factors. On such readings Nazism was a 'perversion, a plague, a catastrophe, and finally a tragedy' that simply befell Germany.[17]

From the 1960s, memory and history in the FRG became more complex and contested as processes of political, generational and disciplinary renewal engendered fresh critical perspectives on the Nazi past. Within this context, and assisted by the growing availability of primary sources, a distinct and mature body of international history writing on Nazi foreign policy emerged.[18] Central to this scholarship was a so-called 'intentionalist' interpretation which identified an ideological programme of expansion, conceived by Hitler in the 1920s and systematically pursued thereafter, aiming successively at the restoration of great power status, pre-eminence in central Europe, continental hegemony and *lebensraum* through conquest of the Soviet Union, imperialist expansion in Africa and the Middle East and finally at some distant date a titanic conflict with the United States for 'world dominion'.[19] These geopolitical ambitions were interlinked with an anti-Semitic and racist agenda to conduct a 'biological revolution' by breeding a 'superior, Germanic elite' – indeed a 'new human being' – and to eradicate the Jewish–Bolshevik archenemy.[20] *Prima facie*, this new explanatory paradigm appeared to be in tune with nascent franker attitudes towards the Nazi past, breaking decisively with post-war obfuscation. It despatched the patently escapist notion that Hitler had been some kind of nihilistic demon, tried to come to terms with the historical evolution and sheer scale of his geopolitical aspirations, and was prepared at least to countenance the possibility that his programme had roots in longer-term German foreign policy traditions. Equally, it had the not inconsiderable merit of bringing his genocidal racism into analytical focus.

When probed for nuance, however, the politics of 'intentionalism' appear rather more perniciously ambiguous. 'Intentionalists' were prepared to admit in fact only a very circumscribed continuity between Hitler and his predecessors, insisting in Andreas Hillgruber's phrase that 'deep breaches and important new departures' were also salient.[21] Klaus Hildebrand accepted that Hitler's continental and overseas ambitions were within the post-Bismarck tradition of German power politics and that his dictatorship was 'the culmination of and surpassed the caesaristic tradition in Prussian-German history'. But set against this were elements of 'revolutionary discontinuity'. Although anti-Semitism was not unknown in German history (hence it functioned well for Hitler as an integrative tool), the fanatical racist dogma that lay at the heart of his project – envisaging genocide, 'world domination' and a biologically engineered 'master race' – was 'new and revolutionary'. Over time, and especially after the beginning of the 'racist war of annihilation' in the east in 1941, the traditional, 'rational', elements in Hitler's thinking began to be undermined and eclipsed by the 'ideological': 'racist dogma had finally triumphed over the political cunning in his Programme'. Thus while 'intentionalism' declined to bracket Hitler off completely, that which was identified as the criminal core of the Nazi project – his 'racist war of extermination' – was categorically presented as 'a "new" feature in the history of Prussia-Germany', and a clear distinction was drawn between his rational, 'pragmatic political calculation' that was in the German tradition and his 'irrational' racism that was a fundamental 'intrusion' (with the former implicitly coded positive).[22]

Thus 'intentionalists' carried over in refined form into the era of would-be objective scholarship the presumption of an exculpatory distance between Nazi racism and mainstream currents in German history, enabling the Third Reich to be accommodated within a still generally positive narrative of the national past. The political valence of this move – a decidedly conservative and nationalist gesture – comes out more clearly when it is contextualized against the other narrative options concurrently available. The rival school of 'functionalist' historians of Nazism prioritized structure and process over ideology and intention, and believed that to stress the singular character of a regime 'devoted to an utterly novel principle for the public order, scientific racism' and to characterize National Socialism primarily in terms of 'Hitlerism' risked foreclosing key questions of continuity and complicity.[23] Equally, liberal partisans of 'historical social science' who were elaborating the notion of a German *Sonderweg* emphasized profound structural continuities and deformities in German history and rejected the notion that after Auschwitz it could simply become a normal nation.[24] Moreover, the very positing of a dualism between power politics and racism was a key postwar mnemonic manoeuvre of 'ultraconservatives' who 'often claimed that it was precisely this addition of biological racism to a more defensible core of political values that [had] baffled them and contributed to their [eventual] turn away from National Socialism': thus in a sense, 'intentionalism' was merely a historiographical manifestation of this generic conservative response.[25]

Of course, the 'intentionalist' interpretation is not mere wishful thinking or indefensible in the light of the evidence. It remains central to scholarly understandings of Nazi foreign policy, accepted as persuasive by foreign as well as German specialists.[26] Moreover, 'intentionalists' do not occupy some completely *outré* position on the political spectrum: leading international historians have played an important role in rebutting the ideologically toxic 'new right' argument that Hitler's war in the east was a defensive response to Bolshevik Asiatic barbarism.[27] Equally, it may well be that there is no way of striking a balance between continuity and discontinuity in explaining Nazism that does not have some problematic implications (and this critique does not imply that there was absolutely nothing novel about Nazi racism). Yet none of this makes reading off the politics of particular formulations illegitimate. As the complex process of negotiating Nazism historically has continued in recent decades, and the centrality of racism to all aspects of its policy and practice has been established, reluctance to regard that racism as an extraneous interpellation into German history has grown, throwing the nationalist connotations of 'intentionalism' into even starker relief.

Indeed, 'intentionalism' as an interpretive paradigm arguably has an inherent problem coping with the current salience of race. A recent account of Nazi foreign policy by Christian Leitz, clearly in this historiographical tradition, is written explicitly against the allegedly disturbing tendency for the Holocaust to dominate the scholarly terrain, and sets out to demonstrate how from 1933 to 1941 foreign policy 'outweighed everything else in Hitler's mind, including deliberation on racial matters'.[28] This insistence upon the separability of power politics and racism must on some level be attributable to a continued historicist attachment to the autonomy and distinctiveness of foreign policy classically defined. Yet in the contemporary landscape of memory, there is something not merely old fashioned but ethically troubling about an account of Nazi expansionism that thus privileges *realpolitik* while almost bracketing off the racial.

In Italy, the nature of the post-war memory settlement has also been the subject of controversial reassessment. With escalating intensity since the end of the Cold War, voices on the right have decried the alleged dominance of anti-Fascist myths during what Prime Minister Silvio Berlusconi in 1994 termed 'fifty years of Marxist hegemony'.[29] Critical scholarship has riposted that the real myth is the notion that anti-Fascism (or Marxism) ever secured such a position of supremacy.[30] Although the post-war republican constitution proscribed the Fascist party and enshrined the resistance ideals of democracy, equality and the rule of law, relatively minimal purging and considerable continuity in state bureaucracy, police and judiciary meant that the break with the past was quite incomplete. Moreover, with unsubtle interference from the United States in the shadow of the early Cold War, the settlement congealed into a paternalistic political order dominated for decades by Christian Democrats and increasingly scarred by clientelism and corruption.[31] Although the left, and pre-eminently the communists, attempted to guard the resistance heritage and retained some influence in the worlds of politics and culture, anti-communism was a far more pervasive force in the early republic than anti-Fascism.[32] Collective memory abetted this conservative reconstruction. The overriding sentiment was to forget the dictatorship and its disasters and to promote a series of dissociating and exculpatory myths. These presented Italians as unwilling victims of Fascism and the war as entirely the work of Mussolini and the Germans, playing on the venerable stereotype of Italians as *brava gente*, fundamentally decent if unheroic folk. Many leftist and liberal anti-Fascists colluded in this 'largely self-absolving collective memory', initially sensing its utility in the manoeuvres to avoid a punitive peace treaty.[33] Thus popular support for Fascism and its criminal brutality were long whitewashed and it is only very recently that the racist oppression of Fascist colonialism, war crimes committed by Italian occupation forces, and complicity in anti-Jewish repression and the Holocaust have begun to emerge from the 'hidden pages of contemporary Italian history'.[34]

Historical representations ploughed similar furrows, eliding both continuity and complicity by presenting Fascism as a European malaise, a parenthesis in Italian national history and a confidence trick perpetrated through threats and mystification by a criminal clique. A dominant discourse of derision belittled it as a historical nullity – with 'neither a vitality of its own, nor an ideology, nor mass support' – and Mussolini as a combination of international gangster and preposterous buffoon.[35] On this view, Mussolini's foreign policy had been tortuously inconsistent, aimless and opportunistic, mixing the pursuit of empty propaganda victories with incoherent outbursts of violence. As prominent liberal anti-Fascist Gaetano Salvemini phrased it, Mussolini was 'always an irresponsible improviser, half madman, half criminal, gifted only – but to the highest degree – in the arts of "propaganda" and mystification'.[36] The visceral appeal of this kind of contemptuous denunciation for triumphant liberal anti-Fascists is self-evident, but in its refusal to take Fascism seriously as a historical phenomenon and its scapegoating of Mussolini it performed an act of distancing that also made it agreeable to mainstream conservative opinion. Indeed, the main dissent from this consensus came from Mussolini's defenders on the far right, such as former propagandist Luigi Villari who lionized him as a staunch anti-communist nationalist and displaced blame for the war elsewhere. For Villari, Mussolini had sought only peaceful revision of the Versailles *diktat* but the British and the French refused to work constructively to this end or to restrain Hitler. Their unreasonable attitude meant Mussolini was ultimately 'forced into the Axis' and then compelled to enter the war by the economic strangulation of British sanctions.[37] (It is perhaps testament to the very success of efforts to minimize the gravity of Fascism that such unrepentant accounts were free to appear.) Anglo-American authors also partook of the dominant mode of representation, portraying Mussolini as a mere 'artist in propaganda', devoid of any principles in foreign policy – here anti-Fascist sentiment was perhaps often spiced with an essentialist disdain for the modern heirs of Roman civilization.[38]

A new phase of scholarship opened from the 1960s, and historians seeking to engage Fascism's historical significance began to accord it the dignity of at least a certain substance, redefining it as a 'political movement driven by clear yet often competing political and economic interests, rather than an exercise in mere political eclecticism, opportunism or chicanery'.[39] As regards foreign relations, historians progressively detected an underlying substance and direction – 'the much stronger presence of elements of planning and a purposive will . . . than Salvemini had supposed' – and some consistent revisionist and imperialist goals.[40] While understandings and knowledge thickened, however, it took historians some time to impose an overall interpretive shape on the subject, and the effort was very much conditioned by the disputatious course of Italian politics in the 1970s. As the left surged closer to achieving national power, laudatory readings of the resistance and critical representations of Fascism gained much greater currency and it sought to entrench them as the ethical and political underpinning of the republic; conversely, conservatives uneasy at the leftward drift of politics and society opposed this nascent anti-Fascist turn in collective memory, attacking the validity of the left's historical vision and thus its claims to legitimacy. In time distinct interpretive positions on Fascism emerged. Anti-Fascist historians increasingly stressed the 'murderousness' of a reactionary regime that they declared the 'revelation' of national history, while more conservative voices sought ways to present the Fascist episode as something other than an unrelievedly negative chapter in the national story.[41] Italian international historians tended overwhelmingly to be in the latter camp. Diplomatic history had been promoted under Fascism and its practitioners had generally assumed that their work 'could and should be harnessed to drive the nationalist dynamo of

Fascist foreign policy'.[42] Little changed post-war, when international historians did not just do ideological work through the production of patriotic historiography, but were often more literally in state service, combining university appointments, curatorial work in state archives and the editing of official documentary collections.[43]

The key figure shaping Italian views on Fascist foreign policy was Renzo De Felice, author of a sprawling and copiously documented multi-volume biography of Mussolini that appeared between 1965 and 1997. This work vaunted its alleged objectivity but outraged the left, since it portrayed Mussolini as a serious far-sighted thinker committed to a third-way modernizing project, and Fascism as embodying an authentic revolutionary drive and winning the broad consent of the Italian people by the later 1930s. Moreover, where the left on discovering something authentic about Fascist ideology had begun to stress its kinship with Nazism, De Felice insisted on a 'fundamental difference' between the 'vitalistic optimism' of the progressive Italian creed and the pessimistic, retrogressive force of the German one. This also impacted on his portrayal of Fascist foreign policy. The Axis, he insisted, was not a product of 'a presumed affinity or, even worse, an ideological identity', but was a tactical manoeuvre. Mussolini had sought to practise the traditional Italian foreign policy of the 'decisive weight', exploiting a position between European power blocs to secure maximum advantage in the shape of colonial gains and a predominant position in the Mediterranean. It was only Western anti-Fascism in the later 1930s that made it impossible to follow a 'pendulum' policy and pushed him into an alignment with a Germany of whom he was actually 'suspicious and fearful'.[44] Even when he entered the war in 1940 it was an 'almost platonic taking up of arms' designed to secure for him the position of 'arbiter *super partes*'.[45] In this comprehensive revisioning, Mussolini had only ever entertained modest expansionist ambitions, was a *realpolitiker* in line with national traditions, and bore less responsibility for the Axis and war than did Britain and France. Where conservative opinion had formerly sidelined and scapegoated Mussolini, De Felice boldly cast him in a positive light as a traditional Italian statesman (in fact resuscitating some themes expounded by Villari) and thus challenged all the shibboleths of contemporary anti-Fascism. De Felice's view proved enormously congenial to Italian international historians, who took up his arguments and subsequently entrenched them as orthodoxy.

Once again, locating De Felice's argument in broader contexts points up its political significance. After the tumult of the 1970s, politics and collective memory both shifted back rightwards, such that by the early

1990s one authority referred to the definitive 'eclipse of anti-Fascism'.[46] The right sought to effect a pacification of the past, banalizing Mussolini by sanitizing his crimes and stressing his positive modernizing achievements (as well as impugning the morality of the anti-Fascist resistance) while also advocating the transcendence of historical passions that were no longer relevant. This was intended both to secure the legitimacy of new political factions such as Berlusconi's *Forza Italia* and the 'post-Fascist' National Alliance and to facilitate the revision of the constitution to purge it of 'anti-Fascist biases' and strengthen the executive.[47] The scholarship of De Felice and his followers carried an obvious political charge in this context, even leaving aside his strenuous active dissemination of his findings as Italy's leading public intellectual. Moreover, it is instructive to contrast his Mussolini with the representation dominant in Anglo-American historiography. As historians outside Italy moved beyond post-war caricatures, they preferred to conceive of Mussolini's foreign policy not as traditional *realpolitik* but rather as 'overtly and insistently ideological'.[48] Fascism meant war and Mussolini had by the mid-1920s devised 'an integrated programme premised on the use of force both internally and externally', in which revolutionary transformation at home and revisionist and imperialist expansion abroad were mutually interdependent.[49] Since his extravagant, dogmatic geopolitical vision could only be realized at the expense of Britain and France, the Axis was no chance contingency but rather a product of shared enemies as well as ideological affinity, indeed a matter of common destiny. On this view, Fascism and Nazism were closely comparable and their histories inextricably interlinked. Moreover, brutality had accompanied bellicosity as Mussolini sought to transform Italians into 'a cruel and domineering master race'.[50] Although Anglo-American international historians equally professed allegiance to a robust empiricist method, in insisting on Fascism's dangerous ambition and essential malignancy they implied that the ethical lessons of anti-Fascism still possessed validity and force.

Japan has long been burdened with a reputation as the most relentlessly amnesiac of the former Axis powers. Former Prime Minister Junichiro Koizumi's refusal to desist from visiting the Yasukuni shrine where major war criminals are interred, and his government's approval of school history textbooks downplaying Japanese aggression and war crimes, have only reinforced the impression of a nation still singularly unwilling to accept responsibility for its wartime actions.[51] Even if recent harsher verdicts on the two comparators rather muddy the waters, there is still truth in this judgement. Although possibilities for radical reconstruction were widely canvassed in the immediate post-war period, ultimately the exigencies of the Cold War and American security requirements ensured that the settlement took on a very conservative cast, symbolized by the almost unbroken political dominance for six decades of the Liberal Democratic Party. Prevailing sentiments towards the wartime past accordingly prioritized Japanese victimization (at the hands of militarist conspirators, Allied bombing and post-war hardship), denial of responsibility for causing the war, sanitization of war crimes and evocations of moral equivalence with the victors.[52] The fact that even in the mid-1990s efforts to concert official proclamations of remorse for the suffering caused by Japan to her Asian neighbours – even in very emasculated and equivocal terms – could spark off heated polemics indicates the deep-rooted nature of these attitudes.[53] That said, the fact that there have always been 'voices of conscience' expressing critical views at odds with official orthodoxy should not be ignored.[54] Progressive historians played an important role for several decades, advocating profound critiques of the structural deformities that had produced Japanese 'fascism', even if they tended 'to be confined in a gilded and permanent opposition' with limited influence on a broader public.[55] Moreover, recent years have witnessed increasing public pressure for recognition and compensation from civil society activists representing women forced into military prostitution and (other) slave labourers. When contemporary Western observers focus on 'the most inflammatory right-wing utterances and interpret them as representative of deep trends', they perhaps overlook the fact that the hysterical tone of such pronouncements derives from paranoia that unpatriotic views of Japan's wartime past are becoming more firmly entrenched in public consciousness.[56]

Conservative arguments concerning pre-war Japanese foreign policy in circulation since the 1940s range across a spectrum, variously minimizing, relativizing or justifying aggression. In some instances, this means endorsing the essential verdict of the Tokyo war crimes tribunal, namely that a conspiratorial militarist clique seized hold of the reins of power and drove the nation into a calculated war of aggression. Thus a late 1940s history textbook firmly stated: 'The Japanese people suffered terribly from the long war. Military leaders suppressed the people, launched a stupid war, and caused this disaster.'[57] On the surface such an argument offers a 'stark narrative of culpability', but actually it allocates blame in a highly restricted fashion, reinscribing the innocence of the mass of the nation.[58] Another variant makes use of the lexicon of exculpation generated by the Tokyo defendants, who certainly did not accept that they had been engaged in a 20-year conspiracy to commit aggression. Rather, 'they believed to the end with all apparent sincerity that their policies, however disastrous in outcome, had been motivated by legitimate concerns for Japan's essential rights and interests on the Asian continent'. Hence they emphasized the threats posed by political chaos and trade boycotts in China, a menacing Soviet Union and communist subversion in Asia, American and European trade protectionism, global trends towards economic autarchy and the coercive encirclement practised by the West prior to Pearl Harbour. The subtext to these arguments was that Japan had not behaved any differently or any worse than other great powers, given the context of an international order collapsing into chaos and brutality.[59] This argument shades into the most extreme nationalist position, which argues that the Japanese were not only operating in legitimate self-defence but actually enacting 'a genuinely moral campaign to liberate all Asia from the oppressive Europeans and Americans, and to simultaneously create an impregnable bulwark against the rising tide of Communism', making the war not only inescapable but altruistic.[60] Although this argument harks straight back to the self-justifications of wartime propaganda, responsible conservative politicians have repeatedly claimed that 'the people of Asia, long colonized by whites, needed to be liberated to give them stable livelihoods' and pondered rhetorically: 'doesn't it take two to wage a war, that is, mutual use of aggression?'.[61]

Positioning themselves against these various competing discourses, mainstream international historians produced a sophisticated and subtly modulated nationalist apologia. The key landmark in post-war scholarship was a seven-volume collaborative project entitled *The Road to the Pacific War*, published in 1962–3 and recognized as path-breaking for its detailed exposition and grounding in unprecedented archival resources (to which the contributors were given privileged access). It has been profoundly influential, being republished in Japanese in 1987 and substantially translated into a five-volume English edition. However, it has also attracted considerable criticism, pointing up how choices of subject matter, focus and enframement can shape the political valence of a text just as much as its overt emplotment. The very decision to give prominence to the Pacific War rather than the conflict in Asia is freighted with ideological significance, since it fits with an exculpatory 'balanced moral calculus' that sets the attack on Pearl Harbour against the atomic bombings, and produces a narrative climaxing with these supreme instances of Japanese suffering.[62] Moreover, it tends to downplay as mere prologue Japanese imperialist aggression in Asia from 1931, and to concentrate instead on the Japanese–American confrontation which is much easier to render either as an inevitable clash of rival imperialisms or a tragedy of misunderstanding. (This latter reading also suited the purposes of liberals keen to find a past that would facilitate contemporary good relations with the United States, which remained central in post-war Japanese foreign policy.)[63] Moreover, this series was explicitly written against Marxist structural critiques of the deeper roots of Japanese aggression, strictly adhering to 'the orthodox approach to diplomatic and military history' and focusing on the minutiae of decision-making at an elite level to the exclusion of broad theoretical or interpretive considerations (and, indeed, associated jargon terms such as 'imperialism').[64] This kind of micro-reconstruction is typically rather depoliticizing in its marginalization of ideological motivations, cultural context and profound forces, and in this instance the approach also disaggregated what Marxists claimed was a coherent, imperialist, 'Fifteen Year War' into a long succession of isolated incidents. Finally, in focusing heavily on the armed forces as actors, the series often invoked a militarist conspiracy thesis echoing that of the Tokyo tribunals, transferring responsibility onto 'precipitate and irresponsible military officers' and stressing 'the importance of military autonomy at the periphery of the Japanese empire'.[65]

Many of these points are evident in the final volume in the series, authored by Jun Tsunoda, who served in the 1930s as an aide to Prime Minister Prince Fumimaro Konoe and was active in nationalist think tanks. Tsunoda offered no critique of Japan's self-avowed mission in Asia, though he amply criticized the United States for a Sinophilia grounded in 'sentimentality, idealism, ignorance, and self-interest'. In densely reconstructing the evolving final confrontation between Japan and the United States, he transferred ultimate responsibility onto American policymakers, who missed opportunities to secure continuing peace through unwillingness to make reasonable concessions; simultaneously, downplaying the deep nationalist and militarist ideological drives animating Japanese policy, he exaggerated the reservations about precipitating conflict that existed in Tokyo and stressed how failures of communication prevented these from exerting a decisive influence and averting the war.[66] On this evidence, it is easy to appreciate the criticism that this series embodies 'an unmistakable effort to shift war responsibility away from Japan'.[67] Although in recent decades new approaches have begun to 'recast the historiography of Japan's diverse diplomacy during the war', this work has canonical status and is still hailed as authoritative.[68] Hence international historians have certainly played their part in determining that 'the single most important problem of "postwar" Japan is [an] inability to come to terms, once and for all, with the pre-1945 past'.[69]

This chapter has explored the ways in which international historians in each of the defeated Axis powers devised explanations of their nation's pre-war aggression that were compatible with – and contributed to – the reconstruction of broadly conservative senses of national identity. Certain structural factors and explanatory tropes recur in each instance – such as the self-pitying rhetoric of victimhood – but there are also individual nuances and points of contrast; for example, it is interesting that in the German historiography continuity arguments are generally coded as frank and critical while in the Italian context they are exculpatory. Collectively, these case studies might seem to suggest that international history as a discourse, with its particular methodologies and preoccupations, is inherently doomed to generate stories palatable to conservative politics. This would be too sweeping a verdict; after all, in the Italian case foreign historians plying precisely the same method as the De Feliceans have produced interpretations purveying a staunchly anti-Fascist message, and in the FRG in the 1960s an arch methodological conservative, Fritz Fischer, overturned a nationalist orthodoxy on the origins of the First World War. Yet there must be something about the particular conjuncture of professional sociology, personal positioning and intellectual preference in each of these cases that led these historians into an unhealthy intimacy with nationalist discourse. At the very least, international historians might be prompted to reflect here on whether their dominant self-understanding as producers of objective scholarly knowledge, doing ideological work only when in error, is entirely adequate. More generally, this material might serve as a partial corrective to the self-congratulatory delusion shared by many historians that our work necessarily functions as an antidote to the fantasies of collective memory and nationalist myth mongering. Neither the conventions and procedures of a professional discipline nor a formidable array of archival sources provide a safeguard against implication in ideological contestation, and if history is not totally in thrall to broader memory discourses it is usually at the very least problematically entwined with them. In these cases, historians exercised enormous ingenuity and dexterity to devise narratives that suited their political purposes. Even after catastrophic defeat and the commission of the most heinous aggressive infamies, it seems that there is very little historians cannot explain away to permit the continued telling of comforting stories about the national us.

Notes

1. Arthur Keith (ed.), *Speeches and Documents on International Affairs, 1918–1937: Vol. I* (London: Humphrey Milford, 1938), p. 50.

2. Annika Mombauer, *The Origins of the First World War* (London: Longman, 2002), pp. 21–118.
3. Wolfgang Schivelbusch, *The Culture of Defeat: On National Trauma, Mourning, and Recovery* (London: Granta, 2004, pb. edn), p. 3.
4. Donald Bloxham, *Genocide on Trial: War Crimes Trials and the Formation of Holocaust History and Memory* (Oxford: Oxford University Press, 2001), pp. 129–53; John Dower, *Embracing Defeat: Japan in the Aftermath of World War II* (London: Penguin, 2000, pb. edn), pp. 443–84.
5. Schivelbusch, *Culture of Defeat*, pp. 1–35.
6. Pieter Lagrou, *The Legacy of Nazi Occupation: Patriotic Memory and National Recovery in Western Europe, 1945–1965* (Cambridge: Cambridge University Press, 2000), p. 305.
7. These cases are discussed more extensively in Patrick Finney, *Remembering the Road to World War II: International History, Collective Memory, National Identity* (forthcoming).
8. Ian Buruma, *The Wages of Guilt: Memories of War in Germany and Japan* (London: Vintage, 1995, pb. edn), p. 8.
9. Robert G. Moeller, *War Stories: The Search for a Usable Past in the Federal Republic of Germany* (Berkeley: University of California Press, 2001), p. 25, quoting Konrad Adenauer.
10. Entry for 5 May 1945 in Anonymous, *A Woman in Berlin: Diary 20 April 1945 to 22 June 1945* (London: Virago, 2005), p. 155.
11. Bloxham, *Genocide on Trial*, underlines how a failure of Allied legal, political and historical imagination also played a major part in this.
12. Friedrich Meinecke, *The German Catastrophe: Reflections and Recollections* (Boston: Beacon, 1963, pb. edn), pp. 93, 1.
13. Jeffrey Herf, *Divided Memory: The Nazi Past in the Two Germanys* (Cambridge, MA: Harvard University Press, 1997), pp. 6–7.
14. Stefan Berger, *The Search for Normality: National Identity and Historical Consciousness in Germany since 1800* (Oxford: Berghahn, 1997), pp. 38, 41.
15. Édouard Husson, *Comprendre Hitler et la Shoah. Les Historiens de la République Fédérale d'Allemagne et l'Identité Allemande depuis 1949* (Paris: Presses Universitaires de France, 2000), pp. 29–47.
16. Ludwig Dehio, *Germany and World Politics in the Twentieth Century* (London: Chatto and Windus, 1959), quotes at pp. 12, 32.
17. Jeffrey K. Olick, *In the House of the Hangman: The Agonies of German Defeat, 1943–1949* (Chicago: University of Chicago Press, 2005), p. 161.
18. Réné Schwok, *Interprétations de la Politique Étrangère de Hitler. Une Analyse de l'Historiographie* (Paris: Presses Universitaires de France, 1987), pp. 83–103.
19. Andreas Hillgruber, *Germany and the Two World Wars* (Cambridge, MA: Harvard University Press, 1981), p. 50.
20. Klaus Hildebrand, *The Foreign Policy of the Third Reich* (Berkeley: University of California Press, 1973), pp. 97, 21–2.
21. Hillgruber, *Germany and the Two World Wars*, p. 41.
22. Hildebrand, *Foreign Policy*, quotes at pp. 144, 135–6, 106–7, 126, 97, 114.
23. Tim Mason, 'Intention and Explanation: A Current Controversy about the Interpretation of National Socialism' in Jane Caplan (ed.), *Nazism, Fascism and the Working Class: Essays by Tim Mason* (Cambridge: Cambridge University Press, 1995), p. 218.

24. Jürgen Kocka, 'German History before Hitler: The Debate about the German *Sonderweg*', *Journal of Contemporary History*, vol. 23, no. 1 (1988), 3–16.
25. Olick, *House of the Hangman*, p. 174.
26. Militärgeschichtliches Forschungsamt (ed.), *Germany and the Second World War. Vol. I: The Build-up of German Aggression* (Oxford: Oxford University Press, 1990).
27. For example, Militärgeschichtliches Forschungsamt (ed.), *Germany and the Second World War. Vol. IV: The Attack on the Soviet Union* (Oxford: Oxford University Press, 1998), though even here the preventative war view is given too much credence considering its origins in Nazi propaganda.
28. Christian Leitz, *Nazi Foreign Policy, 1933–1941: The Road to Global War* (London: Routledge, 2004), p. 5. Cf. Richard Bessel, *Nazism and War* (London: Phoenix, 2005, pb. edn), a text of comparable length with some shared assumptions which fuses the two analysands in its insistence that at the core of Nazism was 'racially conceived struggle and war' (p. 5).
29. Alessandro Portelli, *The Order Has Been Carried Out: History, Memory, and Meaning of a Nazi Massacre in Rome* (Basingstoke: Palgrave Macmillan, 2004), p. 250.
30. Richard Bosworth and Patrizia Dogliani (eds), *Italian Fascism: History, Memory and Representation* (London: Macmillan, 1999).
31. Christopher Duggan, 'Italy in the Cold War Years and the Legacy of Fascism' in Christopher Duggan and Christopher Wagstaff (eds), *Italy in the Cold War: Politics, Culture and Society, 1948–58* (Oxford: Berg, 1995), pp. 1–24.
32. Donald Sassoon, 'Italy after Fascism: The Predicament of Dominant Narratives' in Richard Bessel and Dirk Schumann (eds), *Life after Death: Approaches to a Cultural and Social History of Europe during the 1940s and 1950s* (Cambridge: Cambridge University Press, 2003), pp. 265–90.
33. Filippo Focardi, 'Reshaping the Past: Collective Memory and the Second World War in Italy, 1945–1955' in Dominik Geppert (ed.), *The Postwar Challenge: Cultural, Social, and Political Change in Western Europe, 1945–58* (Oxford: Oxford University Press, 2003), pp. 41–63, quote at p. 47.
34. 'The Hidden Pages of Contemporary Italian History: War Crimes, War Guilt and Collective Memory', theme issue, *Journal of Modern Italian Studies*, vol. 9, no. 3 (2004), 269–362.
35. Richard Bosworth, *The Italian Dictatorship: Problems and Perspectives in the Interpretation of Mussolini and Fascism* (London: Arnold, 1998), pp. 37–81; quote from Emilio Gentile, 'Fascism in Italian Historiography: In Search of an Individual Historical Identity', *Journal of Contemporary History*, vol. 21, no. 2 (1986), 180.
36. Gaetano Salvemini, *Prelude to World War II* (London: Gollancz, 1953), quote at p. 10.
37. Luigi Villari, *The Liberation of Italy, 1943–1947* (Appleton: WI, Nelson, 1959), quote at p. xvii.
38. Denis Mack Smith, 'Mussolini, Artist in Propaganda', *History Today*, vol. 9, no. 4 (1959), 223–32; Alan Cassels, 'Switching Partners: Italy in A. J. P. Taylor's *Origins of the Second World War*' in Gordon Martel (ed.), *The Origins of the Second World War Reconsidered: The A. J. P. Taylor Debate after Twenty-Five Years* (London: Unwin Hyman, 1986), pp. 73–4.

39. John Davis, 'Modern Italy – Changing Historical Perspectives since 1945' in Michael Bentley (ed.), *Companion to Historiography* (London: Routledge, 1997), p. 602.
40. Jens Petersen, 'La politica estera del fascismo come problema storiografico' in Renzo De Felice (ed.), *L'Italia fra Tedeschi e Alleati. La Politica Estera Fascista e la Seconda Guerra Mondiale* (Bolonga: Mulino, 1973), pp. 11–55, quote at p. 24.
41. Bosworth, *Italian Dictatorship*, p. 236.
42. Richard Bosworth, 'Italy's Historians and the Myth of Fascism' in Richard Langhorne (ed.), *Diplomacy and Intelligence during the Second World War* (Cambridge: Cambridge University Press, 1985), p. 89.
43. Bosworth, *Italian Dictatorship*, pp. 82–4.
44. Renzo De Felice, *Fascism: An Informal Introduction to Its Theory and Practice* (New Brunswick, NJ: Transaction, 1976), quotes at pp. 56, 104, 80–2.
45. De Felice, quoted in MacGregor Knox, 'The Fascist Regime, Its Foreign Policy and Its Wars: An "Anti-Anti-Fascist" Orthodoxy?', *Contemporary European History*, vol. 4, no. 3 (1995), 354, 356.
46. Richard Bosworth, *Explaining Auschwitz and Hiroshima: History Writing and the Second World War, 1945–1990* (London: Routledge, 1993), pp. 118–41.
47. David Ellwood, 'The Never-Ending Liberation', *Journal of Modern Italian Studies*, vol. 10, no. 4 (2005), 385–95, quote at p. 393.
48. MacGregor Knox, *Common Destiny: Dictatorship, Foreign Policy, and War in Fascist Italy and Nazi Germany* (Cambridge: Cambridge University Press, 2000), p. 146.
49. MacGregor Knox, 'Fascism: Ideology, Foreign Policy, and War' in Adrian Lyttelton (ed.), *Liberal and Fascist Italy 1900–1945* (Oxford: Oxford University Press, 2002), p. 109.
50. MacGregor Knox, *Mussolini Unleashed, 1939–1941: Politics and Strategy in Fascist Italy's Last War* (Cambridge: Cambridge University Press, 1982), p. 289.
51. Justin McCurry, 'Koizumi Apologises for Wartime Wrongs', *The Guardian*, 16 August 2005, http://www.guardian.co.uk/international/story/0,1549740,00.html (accessed 18 June 2006).
52. John Dower, '"An Aptitude for Being Unloved": War and Memory in Japan' in Omer Bartov, Atina Grossmann and Mary Nolan (eds), *Crimes of War: Guilt and Denial in the Twentieth Century* (New York: New Press, 2002), pp. 217–41.
53. Takashi Yoshida, *The Making of the 'Rape of Nanking': History and Memory in Japan, China, and the United States* (Oxford: Oxford University Press, 2006), pp. 132–4, 141–3.
54. James Orr, *The Victim as Hero: Ideologies of Peace and National Identity in Postwar Japan* (Honolulu: University of Hawai'i Press, 2001), p. 173.
55. Bosworth, *Explaining Auschwitz and Hiroshima*, p. 185.
56. Dower, 'Aptitude', p. 220.
57. Quoted in Saburo Ienaga, *The Pacific War, 1931–1945: A Critical Perspective on Japan's Role in World War II* (New York: Pantheon, 1978), p. 255.
58. Carol Gluck, 'The Past in the Present' in Andrew Gordon (ed.), *Postwar Japan as History* (Berkeley: University of California Press, 1993), p. 83.
59. Dower, *Embracing Defeat*, p. 468.
60. Dower, 'Aptitude', p. 223.

61. Seisuke Okuno and Yoshinobu Shimamura, quoted in Yoshibumi Wakamiya, *The Postwar Conservative View of Asia: How the Political Right Has Delayed Japan's Coming to Terms with Its History of Aggression in Asia* (Tokyo: LTCB International Library Foundation, 1999), pp. 12–13.
62. Gluck, 'Past in the Present', pp. 83–4.
63. Yukiko Koshiro, 'Japan's World and World War II', *Diplomatic History*, vol. 25, no. 3 (2001), 431–2.
64. Sumio Hatano, 'Japan's Foreign Policy, 1931–1945: Historiography' in Sadao Asada (ed.), *Japan and the World, 1853–1952: A Bibliographic Guide to Japanese Scholarship in Foreign Relations* (New York: Columbia University Press, 1989), pp. 222–5.
65. Louise Young, 'Japan at War: History-Writing on the Crisis of the 1930s' in Gordon Martel (ed.), *The Origins of the Second World War Reconsidered: A. J. P. Taylor and the Historians* (London: Routledge, 1999, 2nd edn), pp. 171–2.
66. James Morley (ed.), *The Final Confrontation: Japan's Negotiations with the United States, 1941* (New York: Columbia University Press, 1994), quote at p. 4; note also the critical introduction by David Titus, at pp. xix–xxxviii.
67. Ienaga, *Pacific War*, p. 253.
68. Koshiro, 'Japan's World', p. 435.
69. Naoko Shimazu, 'Popular Representations of the Past: The Case of Postwar Japan', *Journal of Contemporary History*, vol. 38, no. 1 (2003), 116.

8

Defeat, Due Process, and Denial: War Crimes Trials and Nationalist Revisionism in Comparative Perspective

Donald Bloxham

This chapter is concerned with the impact of war crimes trial on defeated polities. It focuses particularly on the aftermath of two cases of criminal warfare in the twentieth century, namely the Ottoman genocide of the Armenians in World War I and the Nazi genocides of World War II, and the war crimes trials conducted correspondingly in Istanbul and western Germany. It addresses the trials as manifestations of what is now fashionably termed 'transitional justice', and thus the way that they influenced the attitudes of implicated peoples to the crimes preceding defeat and 'regime change'. Despite the fundamentally different attitudes displayed by the two states today, the chapter argues that there are interesting points of commonality in terms of their attitudes towards the trials at the time of their enactment. To illustrate and further substantiate some of its arguments, it also refers more briefly to the questions of trying German war criminals after World War I, Japanese war criminals from 1945, perpetrators of ethnic cleansing and related crimes during the post-Cold War break-up of the former Yugoslavia, and perpetrators of the 1994 Rwandan genocide.

In each case, the war criminals question overlaps significantly with broader matters of general, social, moral responsibility for crimes of state, but trials and incarcerations present an interesting prism through which to view the issue. Martin Conway has observed that 'the shape and character of the various purges and prosecutions provide an excellent means of analyzing the dynamics of European societies in the immediate post-war years'.[1] What was true in the post-World War II period is also true for the other situations under examination here. Indeed, the topic is both of historical interest and of contemporary political relevance, given the current investment of the United Nations' political capital and cash in

The Hague, Arusha, and elsewhere as criminal trials have developed into the international community's transitional justice medium of preference, over alternative forms such as the various truth commissions deployed in South Africa and Latin America.

At the levels of psychology and ethics, it is understandable that the international community in whatever constellation should feel it necessary to enact criminal proceedings to make a powerful statement about its outraged values; such, indeed, is the logic that ties together the post-World War I Leipzig trials with the prosecution of 'Hutu Power' members and others. But accepting the 'sense' of trials at those levels does not mean that they will be effective at delivering concrete political outcomes. Firstly, and problematically, the justice delivered in international courts for war crimes and crimes against humanity must inevitably be symbolic for all except the comparatively few perpetrators brought to book, since by dint of sheer practicality the courts cannot reach all of the myriad guilty parties even from within one criminal regime, and by dint of power-political reality some malefactor regimes will be beyond the reach of the courts. Such limits to the reach of the courts will always raise troubling questions of equity and thus to an extent undermine the goal of establishing a general ethical-legal standard, even though they do not undermine the legitimacy of trying those who are within grasp. A second supposed outcome of the existence of international legal machinery is that it deters would-be criminal regimes; no time need be spent on disproving this thesis for the simple reason that its proponents have offered little in the way of evidence to support it. A third consideration, and the main subject of this discussion henceforth, is the impact of trials on the domestic politics of the states whose erstwhile servants and representatives are brought to book.

If the crimes committed in the dissolution of Yugoslavia were the precipitant for the new wave of enthusiasm for international legal action, a point of departure for most optimistic assessments of the role of law in prosecuting state criminality is the Nuremberg trials.[2] 'Nuremberg', for instance, forms an intrinsic part of the logic that predicted reforming Iraqi society after Saddam Hussein's defeat with an occupation regime like that which supposedly transformed Germany into a democracy from 1945.[3] Crudely, since Germany 'turned out OK' in the long run, so the argument goes, 'Nuremberg' and the rest of the occupation must have 'worked'. In considering the social impact of trials, the one relevant historical case about which we have a substantial body of empirical and theoretical scholarship is indeed West Germany in the early post-war decades, but contrary to any idea of Nuremberg as crucible of reform, the trials of the era actually became a focal point of revisionism as nationalist elites successfully pushed for what they pointedly termed the 'final solution of the war criminals question'.[4] The democratization and liberalization that occurred in Germany, as in Japan, did so at first in spite of rather than because of the 'message' of the Allied trials. Yet while 'Nuremberg' confounds more optimistic prognoses for the role of criminal justice in transitional periods, that and similar cases are no less worthy of study for it. To borrow again from Conway, 'the devising and implementation of the structures of justice constituted a highly politicized arena in which mass pressures and elite concerns were focused with a rare intensity'.[5]

Nuremberg and West German memory politics

However positively the German public as a whole today views the Nuremberg trials, and however enthusiastic many German lawyers today are for innovations such as the International Criminal Court and genuinely universal jurisdiction for war crimes and crimes against humanity,[6] at the time it mattered most, namely in the immediate aftermath of Nazi rule, both the medium and the message of the trial were decisively rejected by the west German populace. The influence of generational change in the Federal Republic of Germany, particularly the youth movement of the 1960s which brought with it a more open approach to 'the crimes of the fathers', is the key factor in understanding Germany's retrospective embracing of 'Nuremberg', not the beneficent and re-educative impact of the trials themselves. In other words, cultural changes have influenced as a by-product the way the legal event is viewed in Germany (and elsewhere); the legal event did not shape the cultural change; to argue otherwise is to confuse cause and effect. Indeed, it remained eminently possible to accept German guilt for aggression and genocide and to reject significant parts of 'Nuremberg' as an exercise in supposed American hypocrisy, as happened during the Vietnam War when evidence of American atrocities dovetailed with suspicions of American imperialism. Telling, too, was the inauguration by the German Green Party of a 'war crimes' tribunal in Nuremberg at the height of the arms race, designed to draw attention to US strategy.[7]

From only a short time after the Nuremberg trials, such attention as Germans did give to the 'guilt question' was shaped, on one hand, by the desire for individual self-exculpation and, on the other, by the sense that whatever Germany had inflicted was balanced if not exceeded by the suffering inflicted on Germans. Particular reference was made to

Allied area bombing, the mass rape of German women by Soviet forces in 1945, the legions of German POWs lost to Soviet prison camps, and the Allied-approved forced expulsion into Germany of perhaps 12 million ethnic Germans from eastern and central Europe from 1944.[8] Within the broader context of negative reaction to the occupation, one factor was specific to the institution of trial, however: detachment.

After the initial excitement at the introduction of legal proceedings against the major German war criminals before the International Military Tribunal (IMT) at Nuremberg in 1945, there was a significant ebbing of interest,[9] just as there was even among the judges.[10] The pattern of attention around the IMT trial is in itself instructive, indicating more about the perceived relevance of each component part of the proceedings as its intrinsic interest value. Thus we only read of interest escalating again when the 22 defendants made their own concluding addresses to the court: the final act before judgement.[11] The period of the attention lapse encompassed much of the substance of the trials: the cross-examinations of the individual defendants, and a good part of the presentation of the Soviet case, which contained the most graphic and extensive evidence on crimes against humanity. Conversely, the part of US Chief Prosecutor Robert H. Jackson's opening speech in which he differentiated between Nazis and the mass of ordinary Germans met with much enthusiasm, as did those closing statements of the individual defendants that defended the German people. Indeed, there was a clamour for more substantial press coverage of the latter.[12]

The only sympathy expressed with any of the 'major war criminals' concerned those whom it was felt were not the highest initiators of Nazi policy. Hence amidst the general satisfaction displayed by the contemporary German public at the equity of the IMT proceedings and judgement, the most oft-voiced reservations concerned the fate of the service chiefs. Many did not feel that a soldier or sailor, no matter how deeply complicit, should share the sentence of the overtly political grouping that had compromised him, hence the frequently made contrast between the IMT acquittals and the death sentences for General Alfred Jodl and Field Marshal Wilhelm Keitel. On its most basic level, the principle of differentiation suggested that a general should be executed by the bullet rather than the rope.[13] In these early responses of evasion of responsibility and identification with military servicemen lay some of the seeds that would grow within a short while to full-blown condemnation of the trials that so many Germans had recently accepted.

In the years following the IMT trial, western German attitudes towards the ongoing Allied trial programmes were increasingly shaped by a revisionist German nationalism. German social and political elites, resentful of the Allied occupation, sought to undermine its moral bases and rehabilitate Germany's name by, among other things, attacking war crimes trials as morally and legally unjustified, and playing ever more heavily on the theme of *German* suffering. That the Allies had had to stretch existing international law to cope with the unprecedented brutality of the Third Reich was exploited to the full.[14] Thus arose the revisionist vocabulary, which was to gain popular currency in Germany, of the *Kriegsschuldige* ('war-guilty') and the *Kriegsverurteilten* ('war-convicted'), rather than of the *Kriegsverbrecher* ('war criminals'). And thus arose also the imperative finally to discredit the trials by overturning the verdicts, or at the very least by securing the freedom of the convicts by pressuring the Allies.

The arguments and aims now were of a different nature to the early popular excuses of ignorance and powerlessness, but they fed off their precursors. The shrewdest move made by the elites was to link the two strands in the identification of all war criminals – aside perhaps from some of the 'political' IMT convicts – with the ethic of service to the state.[15] Service, or 'duty', was equated at the time with obedience to senior orders, and it was the unanimous rejection by the various 'war crimes' courts of this principle as a defence that underpinned much of the opposition to trials.[16] The most emotive opposition predictably occurred in the cases of high-ranking soldiers.[17] In 1952 the *Institut für Demoskopie* enquired of Germans in the western zones which of the following group they considered justly imprisoned, and which unjustly: Field Marshal Albert Kesselring (who had been convicted by a British court), Grand Admiral Karl Dönitz, Albert Speer, Rudolf Hess and Baldur von Schirach (all of whom had been convicted by the IMT). The aggregate of respondents placed the men in that order, with the greatest sympathy thus reserved for the two service chiefs.[18] A reservoir of sympathy had earlier been tapped into for Keitel and Jodl, and it was exploited more and more heavily as the new rhetoric identified all convicts with German soldiers, and increasingly regardless of their crime or the organization to which they belonged.[19]

These contentions underpinned the policy towards the Nazi past that was ultimately adopted by most West German political parties. The revisionist line was much more palatable for the majority of the population too,[20] and as in the formation or re-formation of all national communities, a mythologizing re-write of the past was perhaps inevitable – Nazi genocide could certainly not fit any 'optimistic theory' about the present or future.[21] By 1947 the general impulse in West German society too to 'draw a final line' under the recent past – or at least on the suffering that they had caused, if not that which they felt as losers in the war – was in the ascendant.[22]

By the second half of 1947, the second largest-selling newspaper in the British zone, the Christian Democrat *Westfalenpost*, said of the defendants in the Nuremberg 'Doctor's trial' that they were murderers and 'public torturers', but that 'doubtless the interest of the German people in the trial would have been greater if an objective professional German judge had sat on the bench'.[23] By the end of 1949, on the announcement of the 18-year prison term handed-down by a British court to the prominent Field Marshal Erich von Manstein, August Haussleiter of the Bavarian section of the CDU reflected the further development of popular understandings of 'justice' and victimhood. He suggested that such trials struck the public as 'witchcraft trials' if there was no possibility of punishing under international jurisdiction crimes committed on the invasion of Germany and during the expulsion of Germans from eastern Europe.[24] The discourse was shifted from the subject matter of the war criminals cases to the legitimacy of the trials themselves: from the actions of Germany to the actions of the Allies and, by extension, to German victimhood. This concerted assault on the very idea of trial even succeeded in retrospectively influencing German opinion on the IMT trial. In October 1950, the reactions analysis staff of the US High Commission encountered the greatest shift in German societal attitudes ever recorded to that time. Only 38 per cent of a sample of 2000 people regarded the IMT trial as having been conducted fairly, compared to the 78 per cent registered with that view four years earlier.[25]

The German nationalist elites were gifted a potent lever against Allied trial and re-education policy with the onset of the Cold War. The need to placate these leading Germans, and with them broader national sentiment, in the interests of German allegiance in the burgeoning political conflict with the USSR, led first to a winding-down of the war crimes trials programmes in all western occupation zones in the context of a general easing of occupation policy. Later it resulted in a series of more and less politicized 'sentence reviews' for convicted war criminals that sought to 'solve' the 'war criminals question' by the expedient of releasing them all, mostly prematurely.[26] Far from the legacy of Nuremberg being that of a history and morality lesson to the German public, then, by the 1950s the trials were increasingly being seen by the former prosecuting powers as an embarrassment and an obstacle to be removed.

The titanic efforts of the Nuremberg prosecutors in continuing until 1950 with an increasingly controversial legal venture were greatly compromised by the subsequent collapse of the American legal machinery. The final four war criminals in US custody were released by 1958: the number incarcerated in mid-1953 had been 312, and at the beginning of 1955, 41. (Jails in the erstwhile British zone were empty by 1957 after similar rates of reduction. Those imprisoned in the IMT trial were held under quadripartite authority in Spandau jail, and thus could not be released because of the need for Soviet agreement.)[27] Included in the number released after serving only a few years of life sentences and commuted death sentences were commanders of the *Einsatzgruppen* and senior members of the concentration camp hierarchy. Just as telling in the sorry tale of the failed re-educational mission that was the subsequent Nuremberg trial programme is the fact that the edited highlights of the 12 trials, known after publication in Washington DC from 1949–53 as the 15 'green volumes', were never published in Germany. Finally, rejection of the legal validity of the trials was subtly built in to articles six and seven of the 1952 Bonn Treaty ending the Allied occupation statute.[28]

German and Japanese comparisons

Many of the ploys used by German elites after World War II had actually been honed at the end of the previous world conflict. Then, calls to try the Kaiser and at least 1590 others before Allied courts had boiled down to the German domestic prosecution of 13 cases before the Leipzig Supreme Court from 1921. The Allied concession in this matter arose from increasing disunity, involving on one level the general post-war settlement and particularly the competing claims of anti-German revanchism *versus* the need to prevent the conditions in which Bolshevism might prosper, and on another level a fear of violent German nationalist reaction. Indeed, though the very concept of the international Nuremberg court derived in part from the failed experiment in allowing a state to prosecute its own nationals, the international environments of strained inter-Allied relations and Western anti-Bolshevism from 1917–18 as well as from 1945 create an element of contextual continuity between the two episodes. The most significant continuity for the purposes of this chapter, however, is in the rhetoric and technique of nationalist opponents of trial.

It is of course anachronistic to suggest that the overwhelmingly self-exculpatory German response of the interwar years was a rehearsal for the Nuremberg era, but the pattern of nationalist response to the idea, as they saw it, of the nation being impugned by war crimes trials, is abundantly clear. Prominent among the leaders of the agitation were, again, military leaders and veterans groups as well as disproportionate representations from the nationalist right. Just as on its establishment the later Federal Republic would move swiftly to establish a centralized office to provide the best legal assistance for the inhabitants of Allied jails,[29] in July 1919 the German government created a General Committee for the Defence of Germans before Enemy Courts to lend similar assistance to defendants. (While not officially connected with the regime, the committee was headed by a Foreign Office official.)[30] Just as the Nuremberg convicts were spuriously exonerated with the label of the 'war-convicted' (*Kriegsverurteilten*), so too had they been in the 1920s.[31] Even the larger ethical-political context in which specific war crimes were discussed bears the hallmarks of a certain developing continuity, as the 'war guilt question' – *Die Kriegsschuldfrage*, the name, indeed, of a self-exculpatory periodical issued by the Foreign Office – metamorphosed after World War II into the less qualified 'guilt question' – *Die Schuldfrage*, on which Karl Jaspers's meditation of the same name was one of the relatively few genuine attempts at constructive reflection.[32]

The major difference between the two periods of German nationalist outrage after each world conflict is in the greater influence that the earlier exerted over the actual course of the trial programme while it was in progress. Few jurists in the Leipzig court were unaffected by the highly politicized atmosphere. The vast majority of the hundreds of cases provisionally brought to book were not prosecuted to a conclusion. In the remainder, where acquittals were not adjudicated, startlingly low sentences were dealt out, and their severity was often in inverse proportion to the seniority of the defendants' military ranks. This final factor was in itself vital. Alan Kramer writes that the Leipzig trials 'gave the impression of finding guilty only a few expendable individuals guilty of gross transgressions, while the general, collective crimes of the land army were exonerated'.[33] Here the honour of the military and that of the nation were synonymous; in the interest of both entities the war crimes question had to be favourably solved.

Since accusations tend to invite narcissism or defensiveness, the German publics could after each world war identify themselves against defendants or with them. In the former case, as was the early response of some Germans vis-à-vis the highest-ranked war criminals tried in 1945–6, guilt was projected solely onto the men in the dock, allowing the majority of the population to see the trials as irrelevant to them. In the latter case, as was the overwhelming German attitude both in 1919–21 and by the end of the 1940s, guilt either had to be collectively displaced with such devices as the honourable soldier stereotype or the idea of German victimhood. Neither response had anything to do with the genuine introspection that the trials were supposed to bring about. And both responses were, interestingly, facilitated in the Nuremberg era by the inconsistent Allied approach to issues of collective guilt, including, on one hand, mass 're-education' campaigns under the occupation regime and, on the other, for instance, Jackson's courtroom distinction between Nazis and Germans.[34]

The Japanese case presents an interesting blend of the two responses. Here again, the chief prosecutor opened his case to the International Military Tribunal for the Far East (IMTFE) in 1946 with the assertion that 'we must reach the conclusion that the Japanese people themselves were utterly within the power and forces of these accused, and to such extent were its victims'. This fiction was reinforced by the immunity from prosecution granted to the Emperor, Hirohito, in the interests of forestalling patriotic opposition and resistance, particularly in view of the looming conflict with the Soviet Union. The absolution granted to the national figurehead, and the concomitant implicit heaping of the totality of blame onto the 28 defendants in the dock, meant conveniently that for many Japanese the close of the trial drew a line under the wartime past. At the same time there was scope for an equally collectivist rejection of the trial *in toto*. The reality of occupation, the fact that the trial was clearly a function of military defeat, and therefore of power relations, kept alive a resentful sentiment that it was simply a manifestation of partial, victor's justice, and this was reinforced by the not unjustified assumption that prosecution for military aggression was an attack on allegedly deep Japanese political and cultural traditions.[35]

As in post-1945 West Germany, ostensibly contradictory reactions amounted to the same thing in Japan: acceptance of trial precisely because it was seen as exculpating the majority, or rejection of trial precisely because it was seen as indicting the political culture in which the majority had acted. As, too, in West Germany, the marginalization of issues of war guilt and war criminality did not interfere greatly with the incorporation of Japan into the Western politico-economic system in the Cold War and beyond. Where the Japanese case differs from the German is in the level of the ongoing struggle with the past, in which the Tokyo trial is at the core. If politicized responses have now dichotomized into a leftist sense that the trial did not go far enough in its investigations – either into the acts of the emperor or crimes against Asian civilians under Japanese occupation – and a rightist sense that it distorted history to the detriment of the Japan's name, then for ordinary Japanese the IMTFE remains a taboo topic.[36] That such ambivalence did not interfere with the internal 'democratization' process from 1945 should not obscure the difficulties that it still produces in Japan's external relations with its neighbours and former victims.

Republican Turkey and the re-imagining of the 'Armenian Question'

A more extreme and enduringly successful victory in the memory wars was achieved after the Ottoman Empire's defeat in the First World War. From the Ottoman perspective, that war had been fought as a gamble on the preservation of the empire, Istanbul hoping to use the conflict to reduce foreign political and economic influence in its domains, stem the territorial diminution of the polity and even expand eastwards into Russian territory. As an intrinsic part of this drive to divest itself of its economic and socio-political problems, the government deported and massacred its Armenian Christian population in the course of the war in accordance with a desire to ethnically homogenize its Anatolian heartlands. It believed that the Armenian-inhabited areas of eastern Anatolia would at some point be separated from the empire under Russian pressure, and correspondingly accused the Armenians of a wartime community of interest with their co-religionists on the opposing military side.

Allied victory was accompanied by the capture or flight of leaders of the ruling Committee of Union and Progress (CUP) faction, and its replacement with a quiescent regime in Istanbul. The new regime undertook to try its predecessors for, among other things, the murder of the Armenians. These trials did serve to bring much evidence to light on the genocide and command responsibility for it, but they had a distinct subtext: to locate the blame not just for genocide but for the Ottoman entry into war alongside the central powers within a secret conspiracy entirely contained within CUP ranks; in other words, to exculpate the state as a whole both from genocide and from an ill-advised military venture. If this strategy was similar to that of some military and bureaucratic defendants at Nuremberg, who sought to portray German aggression and criminality as the result of the impositions of a small Nazi elite, the Ottoman government could give extra force to the narrative because it itself was running the trials. Indeed, one of the rationales for holding the trials in the first place was to placate the Allies and hopefully ameliorate retribution against the Ottoman state in the form, first and foremost, of the sort of territorial division the CUP had originally gone to war to forestall. At the same time, the courts, like their contemporary counterparts in Germany, were remarkably lenient on the defendants; the few harsh sentences that were handed down were in the main reserved for those culprits who had escaped and were tried *in absentia*.[37]

Given Anglo-French wartime agreements, there seemed no chance of avoiding a division of Anatolia, and the premature ending of the Turkish trial programme in 1920 was directly related to the dawning realization of this reality, though the Istanbul regime also showed itself adept at exploiting divisions between the greedy former Allies, and at taking advantage of their sloth in forging a peace treaty for the Near East. The most trenchant opposition to the Allies, and indeed to British trial plans, came with the rise of a new and vigorous Turkish nationalist movement in the interior under Mustafa Kemal (Atatürk), a movement that was to supplant the Istanbul government and establish the Ankara-based Turkish republican regime. Kemal rallied the remnants of the Turkish army in 1921–2 to defeat the invading forces of Greece, the latter the willing proxy of a Britain seeking to quell any Turkish opposition to its draconian plans for territorial redistribution in and around Anatolia. He thereby assured Turkish control of the rump of the Ottoman territories in Anatolia.

Notwithstanding the continuities of personnel between Kemal's movement and the wartime CUP leadership, and the fact that they could pressure Britain into giving-up its trial plans, the military and diplomatic success of the nationalists did not mean that Turkish elites instantly made the change of course from identifying themselves against previously convicted war criminals to identifying with them. (Though very swiftly they successfully pressured the Allies into foregoing any plans they may have had for trying Ottoman subjects in their custody, and, indeed, into returning those same prisoners to freedom in Turkey.) This was largely down to political opportunism. Just as Kemal replicated the simultaneous approach of the German SPD and that of the CDU after World War II by playing on the possibility of his country succumbing to Bolshevism – except in this case he flirted out of necessity with the new Russian regime – he also found it convenient in gaining international acceptance of his regime to stress the distance between himself and those CUP leaders convicted *in absentia*, referring on occasion both to the trials and their genocidal subject matter.[38] The shift towards the veneration of prominent CUP leaders as exemplary nationalists that we sometimes still see in present-day Turkey only came later, once Kemal had fully entrenched his regime in power and when the subject of the murder of the Armenians became a total taboo from the later 1920s onwards. One aspect of this much-broader process of state-supported erasure of Armenians from Ottoman history – save insofar as they were to be portrayed as a small but treacherous minority – illustrates much about the political failure of the prosecutions of Ottoman elites. The proceedings of the earlier tribunals had been extensively reported in supplements of the official gazette *Takvim-i Vekayi*; with the entrenchment of Kemalism, these records were destroyed to the extent that no Turkish library now holds a complete set.[39]

Alongside the gradual process of national identification with the former genocidaires instead of against them, the entire picture of World War I that the CUP had brought to the Ottoman Empire also metamorphosed. Thus the Kemalists began more and more closely to identify the World War I experience with the later 1921–2 war against the Greeks. The latter war had always been referred to as the national war of liberation against the British-sponsored Greek army, and had concluded with the infamous 'population transfer' involving the simultaneous purging of Ottoman subjects of the Greek orthodox faith and of Greek Muslims; gradually the former war was also referred to by the same epithet, and with analogous referents. In other words, World War I was now re-presented as a war against the powers and their internal Armenian allies. Accordingly, both wars appeared to be but two halves of the same conflict. The black episode of the first 'half' of the prolonged 'war of independence' – the Armenian genocide of 1915 – was alternately denied or presented as a legitimate act of self-defensive war, its perpetrators again depicted as loyal servants of the state. Unlike in Germany after either world war, however, it was possible to re-present the profound defeat in 1918 as part of a larger victory in securing the borders of the 'new Turkey' and removing – in the Armenians and Greeks – its major 'minority problems'.[40]

Contemporary parallels: Rwanda and the former Yugoslavia

Writing in 2001, Payam Akhavan, former legal adviser to the chief prosecutor of the International Criminal Tribunal for the former Yugoslavia (ICTY), concluded that 'the rules of legitimacy in international relations have so dramatically changed . . . during the 1990s that accountability is arguably a reflection of a new "realism"', and that a 'past view of policy based on principles of justice as naïve and unrealistic has been seriously challenged by the convergence of realities and ideals in postconflict peace building and reconciliation'.[41] By Akhavan's lights, what can supposedly be achieved on the international stage – respect for the rule of international law, deterrence of would-be transgressors – is mirrored on the domestic front, in terms of the delegitimization of former perpetrator ideologies, the re-educative effects of trials for population groups complicit in state criminality, the prevention of movements for vengeance by former victim groups and, ultimately, as a result of these factors, the reintegration of societies formerly torn by conflict. He is joined in his optimism by the ICTY itself and the International Criminal Tribunal for Rwanda (ICTR).[42] The website of the ICTY, for instance, boasts that 'by trying individuals on the basis of their *personal* responsibility, be it direct or indirect, the ICTY *personalizes* guilt. It accordingly shields entire communities from being labelled as collectively responsible for others' suffering . . . This paves the way for the reconciliation process within the war-torn societies of the former Yugoslavia'.[43] These points contain a great deal of unsubstantiated assertion.

At the minimum, trials conducted in the context of defeat will be subject to criticisms from the defendants' side. Since in the interwar period the powers failed to create much in the way of enduring principles or an infrastructure of international criminal law, the claims of politicized innovation – *ex post facto* and *ad hoc* law – were as loud in 1945 as in 1919. But irrespective of the establishment of the 'Nuremberg principles' and, therefore, that key legal matter of precedent, the undeniably selective application of those principles to latter-day human rights abusers means that it will always be possible to scream victors' justice by pointing to unpunished crimes and shielded criminals elsewhere as a simultaneous moral defence and national rallying call. This is an important aspect of the relationship between the international strategic and ethical facets of the trial question, on one hand, and the educative effects of trials, on the other hand, since it touches on the *perceived* legitimacy of the trial and thus its claim to interpretative authority among the most immediately relevant audience. Accordingly, the late Slobodan Milosevic, while on trial at the Hague, did not recognize the legitimacy of his court, in common with a majority of Serb and also Croat jurists, but used his platform nonetheless to play to the Serbian community, or at least that small portion thereof that manifested any interest in its proceedings, just as Hermann Göring played at Nuremberg to the attentive minority of the remaining German *Volksgemeinschaft*. Alongside talk of crimes committed against Serbia, Milosevic raised the spectre of NATO co-responsibility for the degeneration of intercommunal relations in the former Yugoslavia. Indeed, since the outside world was cast in blanket fashion as the vindictive enemy, the Serbian nation – with Milosevic as its self-proclaimed representative – could present itself once more as the heroic victim, continuing the trend that contributed so much to the radicalization of Serbian opinion after the collapse of communism.

At the domestic level, for Serbian onlookers the issues at stake are the familiar ones of national identity and responsibility, and unsurprisingly the early months of the trial witnessed an escalation of aggressive nationalist agitation against the proceedings. Given that all national memories are selective, picking and choosing from the historical record in an attempt to form a 'useable past'[44] – which generally means a past in which the nation in question is depicted in a positive light – then Milosevic's minor rhetorical and tactical courtroom victories were always apt to be singled out from the mass of evidence about state-sponsored atrocity to subvert the whole exercise. At the same time, there were some self-serving commentators who suggested that he alone, or a small circle around him, bore complete responsibility for the campaigns of ethnic cleansing.

These divergent responses, mirroring the two equally regrettable tendencies in post-1945 Germany and Japan either to ignore the trials as irrelevant because they only concerned the men in the dock or to reject them as illegitimate because they impugned everyone, illustrate a tension in the medium of didactic trials. This tension resides in the relationship between proving individual guilt for particular criminal acts and illustrating collective socio-political responsibility among the wider publics of the states of which the individual criminals were representatives or senior servants. If the former alone is proven, what real lessons are there for the wider public, and how is the reconciliation advocated by the ICTY and ICTR therefore to be furthered? If the latter is also seriously attempted, how is the prosecution and tribunal to avoid blanket pronouncements of an essentialist sort, since the courtroom, with its emphasis on black-and-white criminal guilt or innocence, is hardly the forum to sort through the myriad shadings of mass responsibility with which the historian or – *qua* Jaspers's fine differentiations in *Die Schuldfrage* – the philosopher has to deal? This conundrum dogged Leipzig, Nuremberg, and Tokyo, and shows no sign of being resolved today.

Though study of the impact of the ICTR and the ICTY and other courts established to further the purposes of these bodies is in its infancy, the research that has been conducted is not encouraging. The evidence from the former Yugoslavia, as from Rwanda, suggests that where trials mean anything to the wider public, popular attitudes divide in significant part along the same ethno-national lines that provided the cleavages for ethnic cleansing and genocide.[45] This certainly indicates ongoing ethnic enmity, but it may also suggest that for Hutu or Serbs, just as for Germans and Japanese at times in both post-world war periods, there was a public sense of the implicitly collective nature of the charges, and an equally collective response (in this case, one of rejection).

Conclusions

Emile Durkheim once observed that stimulating shared reactions across a body politic in reaction to the transgression of shared ethical values could be a socially constructive impact of criminal trials. Mark Osiel's analysis of the potential for and problems of 'didactic legalism' argues that this intrinsically emotionally oriented proposed function of legal reckoning is problematic in that it pays no attention to reason, to 'public articulation and defense of the moral principles underlying legal rules'. Instead, Osiel contends, the very ameliorative function that trials can fulfil in the transition from violent authoritarian to liberal regimes is in the creation of a forum where debate about the past could occur but under particular (liberal) rules of discourse. According to this view, the benefit of the politico-legal system facilitating such a pluralism of views and the testing of one against the other, without imposing either in authoritarian or unquestioning fashion, would be increasingly self-evident. Yet while not denying the theoretical possibilities for Osiel's constructive 'civil dissensus', historical experience suggests that such has not generally been the outcome when peoples have looked upon the past through the legal prism. Instead, a sort of Durkheimian 'mechanical solidarity' has been furthered, though not in directions of which Durkheim would have approved.[46] Solidarity has been furthered not between 'victim groups' and 'perpetrator groups', however defined, but only within them, as the emotive Israeli responses to the Eichmann trial would illustrate in the former case, and the instances addressed heretofore in this chapter would illustrate in the latter case.

The extent to which criticism or rejection of trials can prevail is partly also a function of the state of relations between victor and vanquished. These are complex and dynamic. In the former Yugoslav territories, peace emerged from imperfect political negotiation and compromise, and the political continuities of the various regimes are correspondingly greater than between Nazi and post-war Germany, for instance. Yet even in the context of unconditional surrender in 1945, a resilient conservative-authoritarian German nationalism was enough to defer full societal confrontation with the genocidal past for a generation, while the 'final solution of the war criminals question' was facilitated by a Western alliance happy to make ethical compromises in the quest for western German allegiance in the Cold War. As the Ottoman case showed, the ultimate triumph in increasingly favourable international conditions of a revanchist, Kemalist ethno-nationalism sharing key characteristics and personnel with its CUP predecessor, meant in very short order the abolition of war crimes courts, the release of those hitherto convicted, an amnesty for all, and a policy of state denial of the murder of the Armenians that still holds sway at the time of writing.

The role that can be played directly by victorious powers, either acting on their own part or in the guise of the 'international community', is open to debate, with two distinct sides to the argument. On the interventionist side, this author has stressed elsewhere the necessity for the imposition of broader didactic frameworks into which 'didactic trials' can fit in order to mitigate self-exculpatory trends in the society of the perpetrators.[47] These frameworks were inadequate in important respects in post-war Germany and are inadequate in the former Yugoslavia and in Rwanda.[48] In the absence or limited presence of time-consuming commitments to consistent and persistent civic and historical education programmes and more-or-less objective news media, bolstered where necessary by economic aid and expensive, long-term investments of money and personnel in a military and/or policing presence sensitive to the issues at stake, trials are pointless as 're-educational' tools. Thus whatever the intrinsic merit of the investigative narrative of crimes against humanity established by the ICTY, the marginalization of this on the ground becomes apparent when we see that the recent history of ethnic cleansing has been left out of many Serbian school history books, exactly as the murder of the Jews was tellingly absent from textbooks in the early years of the Federal Republic of Germany. There is also a carrot to be dangled, for just as German openness to the past was surely influenced by the growing stability and confidence engendered by incorporation in the Western Cold War economic system, Serbian leaders have at least had to pay lip service to the merits of the ICTY as one price for better economic relations in Europe.

On the other hand, we have already hinted that the very association of accusations with a purportedly hostile and biased outside world may delegitimate trials. It is again a matter of debate as to how far the European Union's accession requirements on pluralism and human rights have helped liberalize Turkey's political culture to the extent that leftist and liberal intellectuals have recently raised the spectre of the Armenian question. Equally unclear is how far lobbying in other states and on the international stage by Armenian groups has kept the question of the genocide alive even in Turkey. But it is certainly the case that invocation of the issue by third-party states engenders a backlash by nationalist elites who have always viewed pro-Armenian pressure by Christian states as self-serving and imperialist, seeing it as but an extension of the sort of interference in Ottoman 'internal affairs' that actually served to exacerbate intergroup relations in the lead-up to the 1915 genocide.[49]

Whatever the precise balance of factors, whether the Serbian people, for instance, will ultimately traverse the German road of confrontation with the past, the Turkish road of continued forcible displacement, or the ambivalent 'third way' of Japan remains to be seen. Greater openness may only be possible over the course of decades, as the crimes of the post-1989 era are examined by generations with less of a personal stake in denying, minimizing, or shifting the blame for them. What is apparent, however, is that there can rarely be a legal 'quick fix', a force-pacing of the otherwise gradual process of painful change by which states and peoples address the darkest aspects of their histories – if, indeed, they do so at all. For the historian, though, supposedly transitional legal proceedings remain of intrinsic value, if less for the actual impact they have than for what they tell us about the spirit of the time of their enactment.

Notes

1. Martin Conway, 'Justice in Postwar Belgium: Popular Passions and Political Realities' in Istvan Deak, Jan T. Gross and Tony Judt (eds), *The Politics of Retribution in Europe: World War II and Its Aftermath* (Princeton, NJ: Princeton University Press, 2000), pp. 133–56, here p. 134.
2. Geoffrey Robertson, *Crimes against Humanity* (London: Allen Lane, 1999), pp. 202–3; Jürgen Wilke et al., *Holocaust und NS-Prozesse* (Cologne: Böhlau, 1995); Jürgen Wilke, 'Ein früher Beginn der "Vergangenheitsbewältigung"', *Frankfurter Allgemeine Zeitung*, 15 November 1995; Anne Applebaum, 'Justice in Baghdad', *Washington Post*, 19 October 2005, A21; or Roger Cohen writing in the *New York Times*, 30 April 1995, cited in Peter Maguire, 'The "lessons of Nuremberg"', unpublished manuscript, note 3. See also 'Saddam Hussein's Trial Should Be Televised, Says Amherst College Professor', Amherst, MA, 23 August 2005 (A Scribe Newswire) – Lawrence Douglas, posted at http://www.collegenews.org/x4801.xml.
3. Jeffrey Herf, 'Condi Rice is Wrong about Germany's Werewolves but Right about Iraq', *History News Network*, 1 September 2003 at http://hnn.us/articles/1655.html; Applebaum, 'Justice in Baghdad'.
4. Christa Hoffmann, *Stunden Null? Vergangenheitsbewältigung in Deutschland 1945 und 1989* (Bonn: Bouvier, 1992), p. 10.
5. Conway, 'Justice in Postwar Belgium', p. 134.
6. Douglas, 'Saddam Hussein's Trial Should Be Televised'.
7. Dan Diner, *Verkehrte Welten: Antiamerikanismus in Deutschland: Ein historischer Essay* (Frankfurt am Main: Eichborn, 1993), pp. 141–2, 147–8; Werner Jochmann,

Gesellschaftskrise und Judenfeindschaft in Deutschland 1870–1945 (Hamburg: Hans Christians Verlag, 1998), p. 337.

8. Robert G. Moeller, *War Stories: The Search for a Usable Past in the Federal Republic of Germany* (Berkeley, CA: University of California Press, 2001); Norbert Frei, *Vergangenheitspolitik: Die Anfänge der Bundesrepublik und die NS-Vergangenheit* (Munich: C.H. Beck, 1996).
9. Anna J. Merritt and Richard L. Merritt (eds) *Public Opinion in Occupied Germany: the OMGUS Surveys, 1945–1949* (Chicago, IL: University of Illinois Press, 1970), p. 93; Gollancz papers, Modern Records Centre, Warwick University, UK (hereafter, 'MRC'), MSS.157/3/GE/1/17/6, Land Nordrhein-Westphalia reaction report, July 1946; Mass-Observation Archive, University of Sussex, Brighton, UK, File Report 2424 A, 27 September 1946.
10. H. Montgomery Hyde, *Norman Birkett* (London: Hamish Hamilton,1964), p. 518; Carl Rollyson, *Rebecca West: A Saga of the Century* (London: Hodder and Stoughton, 1995), pp. 214–15.
11. *The OMGUS Surveys*, pp. 121–2. This report revealed a particular decline from February 1946 onwards. OMGUS also recorded a diminishment (within the general decline in attention) in the numbers of people reading trial reports in their entirety. See *The OMGUS Surveys*, 34. For the beginning of the decline, see *New York Times*, 16 December 1945, 2 January 1946.
12. *New York Times*, 2 January 1946; MRC, MSS. 157/3/GE/1/17/6, Land Nordrhein-Westphalia reaction report, September 1946, 6–7, 22–3.
13. A plea supported in the case of Jodl by the American and French judges. British National Archives, Kew, London (hereafter 'NA') FO 945/ 332, CCG-COGA, 10 October 1946. Also NA, FO 946/43, 'German Reactions to the Nuremberg sentences'.
14. Frank Buscher, *The US War Crimes Trial Programme in Germany, 1946–1955* (Westport, CT: Greenwood Press, 1989), pp. 92, 100–101, 109–10, 162–3; Frei, *Vergangenheitspolitik*, passim. See also Alfred Streim, 'Saubere Wehrmacht?' in Hannes Heer and Klaus Naumann (eds), *Vernichtungskrieg: Verbrechen der Wehrmacht 1941 bis 1944* (Hamburg: HIS, 1995), pp. 569–97, here p. 575 on some of the spurious *tu quoque* arguments.
15. Buscher, *The US War Crimes Trial Program*, pp. 126, 163.
16. Bodleian Library, Oxford, UK, Goodhart papers, reel 21, Wright to Goodhart, 5 August 1952. See also Peter Steinbach, 'Nationalsozialistische Gewaltverbrechen in der deutschen Öffentlichkeit nach 1945' in Jürgen Weber and Peter Steinbach (eds), *Vergangenheitsbewältigung durch Strafverfahren? NS-Prozesse in der Bundesrepublik Deutschland* (Munich: Olzog, 1984), pp. 13–39, here pp. 17–18, 21.
17. For example, James F. Tent, *Mission on the Rhine* (Chicago: University of Chicago Press, 1982), p. 92.
18. Elisabeth Noelle and Erich Peter Neumann (eds), *The Germans: Public Opinion Polls, 1947–1966* (Westport, CT: Greenwood Press, 1981), p. 202: 6 per cent of interviewees thought Kesselring was justly imprisoned, 65 per cent thought not; the corresponding figures for Hess were 22 per cent and 43 per cent.
19. Bernd Boll, 'Wehrmacht vor Gericht: Kriegsverbrecherprozesse der Vier Mächte nach 1945', *Geschichte und Gesellschaft*, 24 (1998), 570–94, here 592–3, Frei, *Vergangenheitspolitik*, pp. 268–96.

20. Ulrich Brochhagen, 'Vergangene Vergangenheitsbewältigung', *Mittelweg*, 36 (1992/1993), 145–54, especially p. 149.
21. Jeffrey Herf, *Divided Memory: The Nazi Past in the Two Germanys* (Cambridge, MA: Harvard University Press, 1997), p. 392.
22. Frei, *Vergangenheitspolitik*, p. 14.
23. NA, FO 1056/239, German press review, 15 August to 15 September 1947.
24. For example, *Frankfurter Allgemeine Zeitung* 20 and especially 21 December 1949.
25. Anna J. Merritt and Richard L. Merritt (eds), *Public Opinion in Semisovereign Germany: The HICOG Surveys, 1949–1955* (Chicago, IL: University of Illinois Press, 1980), p. 101.
26. Donald Bloxham, *Genocide on Trial: War Crimes Trials and the Formation of Holocaust History and Memory* (Oxford: Oxford University Press, 2001), ch. 5; Buscher, *The US War Crimes Trial Programme*; Thomas Alan Schwartz, 'Die Begnadigung deutscher Kriegsverbrecher: John J. McCloy und die Häftlinge von Landsberg', *Vierteljahreshefte für Zeitgeschichte*, 38 (1990), 375–414; Peter Maguire, *Law and War: An American Story* (New York, NY: Columbia University Press, 2001). On German rearmament, David Clay Large, *Germans to the Front: West German Rearmament in the Adenauer Era* (Chapel Hill, NC: University of North Carolina Press, 1996).
27. Figures from Maguire, *Law and War*, p. 256.
28. Maguire, *Law and War*, p. 229.
29. Frei, *Vergangenheitspolitik*, pp. 21–2.
30. James F. Willis, *Prologue to Nuremberg* (Westport, CT: Greenwood Press, 1982), p. 113.
31. Jürgen Matthäus, 'The Lessons of Leipzig – Punishing German War Criminals after the First World War', forthcoming in Patricia Heberer and Jürgen Matthäus (eds), *Atrocities on Trial: The Politics of Prosecuting War Crimes in Historical Perspective* (Lincoln, NE: University of Nebraska Press, 2008), for the use of the expression after the First World War.
32. Karl Jaspers, *Die Schuldfrage* (Heidelberg: Lambert Schneider, 1946).
33. Alan Kramer, 'The First Wave of International War Crimes Trials: Istanbul and Leipzig', *European Review*, 14, no. 4 (2006), 441–56, here 450. The most comprehensive study of the whole trial programme is Gerd Hankel, *Die Leipziger Prozesse. Deutsche Kriegsverbrechen und ihre strafrechtliche Verfolgung nach dem ersten Weltkrieg* (Hamburg: Hamburger Edition, 2003).
34. Bloxham, *Genocide on Trial*, ch. 4.
35. Madoka Futamura, 'Individual and Collective Guilt: Post-War Japan and the Tokyo War Crimes Tribunal', *European Review*, 14, no. 4 (2006), 471–84; see also Futamura's contribution to this volume. More generally on the trials and Japanese responses, see John Dower, *Embracing Defeat: Japan in the Aftermath of World War II* (London: Penguin, 2000).
36. Futamura, 'Individual and Collective Guilt'.
37. Willis, *Prologue to Nuremberg*, pp. 153–63; Vahakn N. Dadrian, 'Genocide as a Problem of National and International Law: The World War I Armenian Case and Its Contemporary Legal Ramifications', *Yale Journal of International Law*, 14 (1989), 221–334, especially 278–91. On trials and their broader geopolitical context, see Taner Akçam, *Armeniern und der Völkermord: Die Istanbuler*

Prozesse und die türkische Nationalbewegung (Hamburg: Hamburger Edition, 1996); idem, 'Another History on Sèvres and Lausanne' in Hans-Lukas Kieser and Dominik Schaller (eds), *Der Völkermord an den Armeniern und die Shoah* (Zurich: Chronos, 2003), pp. 218–99; Donald Bloxham, *The Great Game of Genocide: Imperialism, Nationalism, and the Destruction of the Ottoman Armenians* (Oxford: Oxford University Press, 2005), ch. 4.
38. See previous note.
39. Donald Bloxham and Fatma Müge Goçek, 'The Armenian Genocide' in Dan Stone (ed.), *The Historiography of Genocide* (Basingstoke: Palgrave Macmillan, forthcoming, 2008).
40. Bloxham, *The Great Game of Genocide*, pp. 198–9.
41. Payam Akhavan, 'Beyond Impunity: Can International Criminal Justice Prevent Future Atrocities?', *American Journal of International Law*, 95, no. 1 (2001), 7–31, here 30.
42. On the ICTR, see Rosemary Byrne, 'Promises of Peace and Reconciliation: Previewing the Legacy of the International Criminal Tribunal for Rwanda', *European Review*, 14, no. 4 (2006), 485–98.
43. http://www.un.org/icty/glance/index.htm.
44. On the idea of a 'useable past', see Moeller, *War Stories*.
45. Eric Stover and Harvey Weinstein (eds), *My Neighbor, My Enemy: Justice and Community in the Aftermath of Mass Atrocity* (Cambridge: Cambridge University Press, 2004); Anna Uzelac 'Hague Prosecutors Rest Their Case' *Institute for War and Peace Reporting*, 27 December 2004; relatedly, Human Rights Watch, 'Justice at Risk: War Crimes Trials in Croatia, Bosnia and Herzegovina, and Serbia and Montenegro', *Human Rights Watch*, 16, no. 7 (2004), 1–31. Even one of the more positive assessments of the ICTY's impact in Bosnia concludes that its 'main contribution seems to have been its utility as a political lever, rather than its utility as a tool of post-conflict reconciliation'. See Rachel Kerr, 'The Road from Dayton to Brussels? The International Criminal Tribunal for the Former Yugoslavia and the Politics of War Crimes in Bosnia', *European Security*, 14, no. 3 (2005), 319–37, here 331.
46. Mark Osiel, *Mass Atrocity, Collective Memory, and the Law* (New Brunswick, NJ: Transaction, 2000), chs 1 and 2, quotation from p. 31.
47. Bloxham *Genocide on Trial*, pp. 80–8, ch. 4, especially p. 132.
48. On Yugoslavia, Uzelac, Hague Prosecutors rest their Case, p. 7; Kerr, 'The Road from Dayton to Brussels', 324.
49. Bloxham, *The Great Game of Genocide*, ch. 1.

9

Memory of War and War Crimes: Japanese Historical Consciousness and the Tokyo Trial

Madoka Futamura

On 2 September 1945, the Japanese Government signed the Instrument of Surrender and accepted the Potsdam Declaration, which defined terms for Japanese surrender. In its article 10, the Declaration stated that 'stern justice shall be meted out to all war criminals, including those who have visited cruelties upon our prisoners', which became the legal basis for the following trials of Japanese war criminals: the International Military Tribunal for the Far East, the so-called Tokyo Trial, and other trials for minor war crimes. Considering the fact that these trials were an integral part of the Allies' occupation policy to demilitarize and democratize post-war Japan, they need to be assessed within that context.[1]

This chapter, however, focuses on the Tokyo International Military Tribunal and its impact on Japanese historical consciousness with regard to the Asia-Pacific War. This approach derives from the fact that it has been pointed out both by academics and practitioners that war crimes trials play, or are expected to play, some roles in creating post-conflict collective memory.[2] Whether and to what extent such trials influence collective memory is debatable; however, as Mark Osiel argues, war crimes trials have been used 'with a view to teaching a particular interpretation of the country's history': criminal law has been employed to 'cultivate a shared and enduring memory' of mass brutality.[3] Is this also the case with post-war Japan? Having organized the symposium on the Tokyo Trial in 1996, historian Takeshi Igarashi stated: 'our understanding about how much the Tokyo Trial had left a deep scar on the Japanese people's perception of history was not enough.'[4] What is the nature of this scar? This chapter examines whether, and in what way, the Tokyo Tribunal has had an impact on the Japanese people's historical consciousness and collective memory of the past war experience, by looking at how the Tokyo Trial has been perceived and debated by the Japanese.

The Tokyo Trial

The International Military Tribunal for the Far East was established in 1946 to prosecute and punish Japanese wartime leaders for conducting war crimes. It was composed of 11 judges from the nine signatories of the Instrument of Surrender – Australia, Canada, China, France, Great Britain, the Netherlands, New Zealand, the Soviet Union, and the United States – plus India and the Philippines, each one of which also offered a member for the Prosecution section. The Charter of the Tokyo Tribunal was modelled on the Nuremberg Charter and set 'crimes against peace', conventional war crimes, and 'crimes against humanity' as its jurisdiction. Indictments were issued for 28 defendants, which included General Hideki Tōjō, who was Prime Minister at the time of the attack on Pearl Harbour, and 17 other military officers. Four of the defendants were former Prime Ministers, and most others were members of wartime cabinets. The Tokyo Trial opened on 3 May 1946 and lasted until 16 April 1948. The judgement was rendered on 12 November 1948 to 25 defendants, excluding two defendants who had died during the Trial and one who has been discharged because of a mental disorder. All of the 25 defendants were found guilty. Seven, including Tōjō, were sentenced to death, 16 to life imprisonment, one to 20 years' imprisonment, and one to 7 years' imprisonment.[5]

Although it is understood as an 'international' tribunal, the outset of the Tokyo Trial was very much coloured by the United States. The Charter was issued through an executive decree of General MacArthur, who had the authority to appoint the justices as well as the President of the Tribunal.[6] For the United States, the Tokyo Trial was an integral part of its occupation policy to demilitarize and democratize post-war Japan.[7] For these ends, the punishment of war criminals was not enough; the criminality of those individuals as well as the war as a whole had to be accepted by the Japanese. 'The purpose of the trial was to *convince* the Japanese people that their leaders misled them into war', Owen Cunningham, a member of Defence Counsel at the Tokyo Tribunal stated.[8] Establishing the history in the minds of the Japanese was an important part of the 'psychological campaign for demilitarisation',[9] and in this context, not only the verdict but also the Tribunal's detailed account of the war were expected to have pedagogical effects for the vanquished nation.

The judgement of the Tokyo Trial

Through its legal procedures and judgement, the Tokyo Trial provided a historical record of the war and war crimes conducted by Japan. The judgement basically accepted the case of the Prosecutor and concluded that there was, in Japanese policy, a criminal conspiracy to wage wars of aggression against the Allied Powers. The judgement defined the conspirators as the military and their supporters, who 'carried out in succession the attacks necessary to effect their ultimate object that Japan should dominate the Far East'.[10] The judgement also found that the Japanese Army had perpetrated serious war crimes against the Allies' POWs, and against civilians while waging war in China, including the notorious Nanjing massacre.

It is now widely agreed among academics that it is risky to accept unconditionally the account of the war and Japanese wartime policy provided in the judgement. One of the flaws was the Tribunal's application of the concept of conspiracy.[11] The Prosecution at the Tokyo Tribunal insisted that 'it is apparent from the evidence that there did exist a really carefully planned conspiracy or common plan for commission of the crimes set forth in the Indictment'.[12] However, historian Makoto Iokibe posits that, in the case of Japan, the whole war was 'not the matter of the existence of cool-headed and evil "conspiracy" but the matter of its *non-existence*'.[13] The fact that the Japanese cabinet changed 17 times during the period between 1928 and 1945 reveals the complexity and incoherence of wartime Japanese policy during the period. What is more, a staff member in the attorney general's office also stated: 'Japan never had any consistent party like the Nazis, and out of a plethora of possible defendants responsible since 1931, only a few can be picked from a very competitive bunch.'[14] Conspiracy is a convenient concept that enables wholesale prosecution of military leaders; however, it is too simplistic a concept on which to base historical narration.

The second shortcoming of the Tokyo Trial's account of the war is its heavy focus on the role of the military clique. To explain 18 years of Japanese history only by a conspiracy of the military clique is too simplistic. Iokibe claims that the procedures of the Tribunal lacked two important civilian figures who played a significant role in expanding the war: Yōsuke Matsuoka, a former Minister of Foreign Affairs, who concluded the Axis Alliance, and Fumimaro Konoe, a former Prime Minister who contributed to the expansion of the Sino-Japanese war.[15] Shinichi Arai claims that examining the developments which led to the war within the framework of extreme militarist versus moderate political leaders is the 'elitists' view of history', which lacks 'structural analysis of social and historical elements that have enabled Japan to conduct aggression'.[16]

Indeed, the Tokyo Trial exempted several important issues and people from prosecution as a result of political calculations. One example is the Emperor, who was not indicted or even summoned to the Tribunal. The prosecution of the Emperor was closely aligned to the occupation policy of the United States. As General Douglas MacArthur, the Supreme Commander for the Allied Powers in Japan, for example, argued, with Japan becoming a vital strategic ally with the start of the Cold War, the risk of angering the Japanese if they disgraced the Emperor was a concern.[17] In addition, the Emperor was seen to have some utility in the process of re-uniting the confused war-torn society. Nevertheless, the total absence of the Emperor in the record of the war is a fatal blow in terms of clarifying Japan's social structure and its relationship to the war.

The third shortcoming is that the Tokyo Trial, as well as its judgement, focused heavily on the Pacific War. Kentarō Awaya points out that World War II in the Asian-Pacific sphere had three aspects: war among imperial powers, war between fascist and anti-fascist powers, and liberation war against colonialism. The Tribunal, however, created a simple account of the war based on the Allies' view of the war, that is, war between fascist Axis and anti-fascist Allies, rather than examining the complex nature of the war. Accordingly, 'the aspect of the war among the imperial powers was overlooked and the aspect of liberation against colonialism was dealt with subordinately.'[18] Michitaka Kainō, a member of the Defence Counsel at the Tribunal, claimed:

> The Tokyo Trial should have focused more on Japan's war against China. As a war, the Pacific War was bigger in scale and impact, causing greater harm to the world. However, on the point of immorality and aggressiveness, it is doubtful whether Japan's war in the Pacific should be focused on. Something is wrong with regarding the war crimes trial as being a mere Tōjō trial.[19]

The Tokyo Trial was conducted with a specific view of the war. This can be seen from the fact that the Tribunal set its jurisdiction between 1 January 1928 and 2 September 1945, implying that this was *the* period of Japan's war. What is more, such a temporal jurisdiction could and did determine who was the villain of the war. For these reasons, many Japanese academics believe that 'In a real sense . . . the Tokyo trial sat in *judgement on the history* of the Pacific region up to the end of World War II.'[20] In other words, the Trial had defined the nature of the war. John

Pritchard states: 'There is, perhaps, a philosophical divide between those who call it the "Tokyo War *Crimes* Trial" and those who call it the "Tokyo *War* Trial". I prefer the latter.'[21]

Some pointed out a problem of resorting to legal procedure for the account of the events.[22] Even a high official in the then US government expressed some doubts. William Sebald wrote as follows: 'the validity of the trial itself doubtless will be long debated, as will the wisdom of using a court process to reveal this military past to the Japanese.'[23]

In spite of the above shown problems in the accuracy of the Tokyo Tribunal's historical record, it is important to note that the Tokyo Trial's account of the war had a significant impact on the Japanese at the time because it revealed the detailed fact of the war and war crimes, which had been concealed from the public during the war. This information convinced the nation that they had been deceived by their wartime leaders, especially the military. 'During the war we were forced to suffer a poor life; but we lost the war that Tōjō had said that we would definitely win', an executive of a company said during the Tokyo Trial: 'Now I came to learn through the Tokyo Trial and others that it was a reckless, aggressive war pursuing the interests of the privileged class and capitalists, and realized that we had been completely deceived.'[24]

What is important here is that people accepted the Tribunal's account of the war. The acceptance of the Trial's historical account was natural, considering the fact that the Tokyo Trial together with the whole occupation was accepted albeit passively as the inevitable result of defeat.[25] In addition, the GHQ's propaganda campaign before the Tokyo Trial, which included the running of a series of articles, with titles such as 'The History of the Pacific War – The Source Offered by the GHQ', in major newspapers, may also have had a substantial effect.[26] More importantly, the Trial's account of the war was 'comfortable' for the people at the time in several aspects. First, the judgement excluded the role of the Emperor, who was strongly supported by the nation even after the war.[27] Second, by focusing on the role of the military clique, the Tokyo Trial pointed the finger at those who should be blamed for the war, defeat, and misery, and by doing so, it freed the majority of the population from the burden of war guilt. The Japanese accepted the Trial's judgement because it matched the national sentiment at the time: anger towards the wartime leaders and the military, combined with people's disassociation from them. One political cartoonist expressed his hatred for the defendants at the Tokyo Trial: 'All of us people were deceived and used by them, and cooperated in the war without knowing the true facts. Looking back now, this was because of ignorance and being deceived.'[28] The people were also willing to leave the war behind and move forward.[29] The Tokyo Trial thus had a symbolic effect not only by detaching the Japanese from their wartime leaders but also by giving people a process through which to bury the militarist past and start from scratch as a peace-loving, democratic nation. This is an important element in the Japanese seemingly passive acceptance of the Tokyo Trial's judgement and its account of the war.

The 'Tokyo Trial view of history' and the revisionist view

For some people, however, the view that 'Japan has waged wars of aggression and perpetrated terrible war crimes' is scarcely acceptable. Such a view is even more intolerable when it is seen as having been given, or imposed, through an international trial conducted by the victor of the war. The Tokyo Trial's account of wartime Japan has been bitterly criticized by those who claim that the wars Japan fought from the late 1920s until 1945 were not aggressive but self-defensive in nature, or that it was the war to liberate Asia from Western imperialism. The 'Tokyo Trial view of history' (*the Tōkyo saiban shikan*), they argue, was an interpretation imposed by the victor. In the place of the 'Tokyo Trial view of history', these critics are willing to promote an affirmative, or glorifying, account of pre-war and wartime Japan. This is why anti-'Tokyo Trial view of history' critics show stronger resentment of the Tribunal's account of the Japan's war crimes than of its silence over American alleged war crimes: Hiroshima and Nagasaki. The Nanjing massacre first came to be widely known to the Japanese public through the Tokyo Trial, which recorded that the Japanese Imperial Army killed, raped, and tortured more than 200,000 people in the city. For those who refuse to accept the fact of the massacre, the Tokyo Trial is detestable as the origin of the 'fabrication' of the crime.[30]

This kind of claim remained in the minority within the society, until the 1990s. At the beginning of the 1990s, individuals in Asian countries started to raise their voices, asking for an apology and compensation for the suffering inflicted on them by Japan during the Second World War. Facing increasing claims from the victims of Japanese war crimes, notably from the 'comfort women', the Japanese government publicly acknowledged the issue in 1993 and sponsored the establishment of an NGO to provide 'sympathy money' to the victims. Several prime ministers expressed atonement for a war of aggression, and the so-called war apology resolution was passed in 1995 by the National Diet under Prime Minister Tomiichi Murayama of the Social Democratic Party of Japan.[31]

While the government's actions were criticized by some as insufficient, they angered and frustrated some others as too 'apologetic'. What angered the latter was the Japanese people's 'uncritical' acceptance that 'Japan conducted wars of aggression and committed war crimes'. It was in this context that these critics attacked the Tokyo Trial, especially its verdict and historical record of wartime Japan, as the source of such a 'masochistic' attitude. Tōru Maeno argued that many Japanese were losing their pride and identity as the 'Japanese race [*Nihon minzoku*]':

> The guilty verdict at the Tokyo Trial that concluded that Japan was the aggressor totally denied Japanese tradition and culture, which had been established up to then. Japanese history was distorted and the truth of the Greater East Asia War was concealed. Since then, the Japanese have been overcome with a sense of guilt, and many Japanese have come to regard Japanese history and traditional culture negatively based on the 'Tokyo Trial view of history'.[32]

In 1998, a film, *Puraido: Unmei no Toki* (Pride: The Fateful Moment), which focused on the 'heroic agony' of Tōjō at the Tokyo Tribunal, was released.

The so-called neo-nationalist movement in the 1990s led to a series of 'revisionist' movements which attacked not only the 'Tokyo Trial view of history' but also those whom they regard as having proliferated and sustained the 'enemy propaganda' ever since the occupation: journalism, the 'progressive intellectuals', and the school teachers' union.[33] In 1997, an association called the 'Japanese Society for History Textbook Reform' (*Atarashii Kyōkasho wo Tsukurukai*) was established. This association attacked Japanese school history textbooks for depicting the Nanjing massacre and the comfort women, and thus proliferating the 'masochistic' view of history to children. The association published its own textbook, which put forward the view that Japan's war was Asia's liberation war against the Western powers; it caused strong criticism from China and South Korea, and raised heated debates both domestically and internationally.[34]

Such neo-nationalist and revisionist arguments in the 1990s have been widely publicized and also supported by a number of public figures. For example, in 1994, the Minister of Justice stated in public that 'the Nanjing massacre was a fabrication' and 'the war was not an aggressive war'. He was forced to resign for his statement.[35] Neo-nationalist criticism of the Tokyo Tribunal appealed to not a few war bereaved who resisted the idea that the war was an aggression, because to accept that would be to disgrace the soul of their family members and to admit that they died in vain.[36]

General apathy towards the Tokyo Trial: Ignorance, cynicism and taboo

Despite frequent media coverage of the recent Japanese nationalist posture, there has been a strong rejection of the past Japanese war and militarism among the nation.[37] The opinion poll conducted by *Mainichi Shimbun* in August 2005 shows that 43 per cent thought that Japan's wars against the US and China were 'wrong wars', while 29 per cent thought they were 'inevitable wars'.[38] According to another opinion poll conducted by *Asahi Shimbun* in 1994, 72 per cent answered that 'Japan has not done enough to compensate people under the Japanese colonial rule and occupation'.[39] The 'wrongfulness' of the war remains the perception of the majority of the nation, and voices raised against the Tokyo Trial and the 'Tokyo Trial view of history' can be regarded as a nationalist's reaction to such a national perception

However, it is not clear whether this Japanese perception can be attributed to the impact of the Tokyo Trial, as the neo-nationalists and revisionists claim. War crimes trials alone cannot cultivate historical consciousness. What is more, the Tokyo Trial in fact is not visible in Japanese collective memory and public discourse. Rather, the general attitude of the Japanese regarding the Trial has been 'apathy'. The main reason for the apathy is ignorance. According to an opinion poll of *Asahi Shimbun* on 2 May 2006, 70 per cent answered that they did not know the content of the Tokyo Trial well.[40] The author also conducted interviews in 2003 and found that the immediate reaction of Japanese was often one of ignorance and indifference.[41] However, close examination shows that the apathy consists of multiple elements. In addition to 'ignorance', there is 'cynicism' and a sense of 'taboo'.

Among the Japanese, it has been widely held that the Tokyo Trial was 'victor's justice'. 'Victor's justice' caused bitterness and remains as a scar in their collective memory.[42] However, unlike the vocal nationalists and revisionists, the general reaction to 'victor's justice' is a passive acceptance. As noted above, the Tokyo Trial was, and is, accepted as 'inevitable' within the context of defeat. Such an acceptance is often accompanied by cynicism because the Trial has been regarded as 'a result of defeat, nothing more'. Cynicism deprives people of the will to examine the general significance and lessons of the Tokyo Trial as well as issues that were examined there.[43]

This national reaction illustrates the very problem surrounding the Tokyo Trial and the Japanese: that is, the Trial has been either talked about emotionally and ideologically within a limited circle or simply ignored by the majority of the population. And this is strongly related to the third element of the apathy: the general perception that the Tokyo Trial is a national taboo. The Trial is a taboo partly because of what was judged. Aggressive wars and terrible war crimes are not something that the nation can proudly talk about. What is more, for the Japanese, having their own leaders externally and unilaterally punished is an uncomfortable experience. Daizaburō Yui, a historian, pointed out in the late 1990s that a sense of unfairness, still existing among the Japanese, prevented people from accepting the Tokyo Trial unambiguously.[44] In addition, the experience of war crimes trials does not fit in well with Japan's post-war identity as a peace-loving nation. According to some academics, there is 'a radical break in 1945' in the Japanese psyche, that is a clear demarcation between pre-war and post-war Japan.[45] Although the Tokyo Trial was conducted during the occupation as both 'the last act of the war and first act of the peace',[46] to the Japanese, it was much more of 'the last act of the war', which they rather want to push to the bottom of their collective memory together with the war.

The taboo also extends to issues that were *not* judged by the Trial: the Emperor's war responsibility and the alleged crimes of Unit 731, for example. As to the issue of the Emperor's war responsibility, it was not until his death in 1989 that it came to be actively discussed in public. In that year, *Asahi Shimbun* conducted an opinion poll on the Shōwa Emperor, which included a question about whether the Emperor had any kind of responsibility for the Second World War. Twenty-five per cent answered yes and 31 per cent no; however, 38 per cent replied they could not say either yes or no.[47] To examine the Tokyo Trial, the Japanese have to face this delicate issue. In addition, there is a guilty consciousness among the wartime generations. They hesitate to examine the Tokyo Trial because it makes them face their own guilt, a matter which the Trial did not judge.[48] The Japanese, as a nation, supported the war, and people are well aware deep down that the nation's responsibility cannot be decollectivized simply by individual criminal punishment.[49]

The most significant aspect of taboo, however, comes from the fact that the Tokyo Trial is strongly related to another sensitive issue: how to see the nature of the war, a matter of political and ideological debate between the right and left. Indeed, in Japan to this day there is no consensus on how to refer to the war that ended in 1945: 'the Second World War', 'the Asian-Pacific War', 'Fifteen-Years War', or 'the Greater East Asia

War', each of which is strongly associated with different views on the nature of the war.[50] Many of the government's and the Emperor's statements still call the war 'the *previous* war' (*saki no sensō*). This illustrates well that the Japanese have not yet completed a summary of the war.

The Tokyo Trial is related to this unsettled business because the Trial itself offered a specific account of the war. This fact makes the Trial a difficult topic to discuss. Being critical of the Tokyo Trial does not necessarily mean denying the Tribunal's judgement that Japan conducted aggressive war and war crimes, and valuing the Trial is not equal to the acceptance of the whole judgement. However, they are inseparable in the Japanese mentality. The connection between the Tokyo Trial and its account of the war is so strong that it is not possible to criticize the former without being seen to challenge the latter, and doing so has been a taboo in post-war Japan. On the other hand, the Tokyo Trial has been criticized by those who try to justify, or glorify, Japan's past war. In their argument, it is not always clear whether their claim about a Japanese 'masochistic attitude' refers to the acceptance of the fact that Japan conducted an aggressive war, or to the acceptance of the historical account that was imposed by victor's justice. The inaccuracy of a certain view of the war and the inappropriateness of the way the view was proposed are two different issues, but they were mixed-up in an emotional argument that aims at the total negation of the Tokyo Trial.

What is more, neither the right nor left is satisfied with the Tokyo Trial's historical account. The rightist critique cannot accept the Tokyo Trial for having conveyed the view that has made the Japanese obsequious and masochistic, and the leftist view criticizes the Tokyo Trial for not having done enough to reveal Japan's war crimes against Asian civilians and the Emperor's responsibility. Either way, it is difficult for the Japanese to examine the Tokyo Trial neutrally and objectively. This is linked to the fact that the Tokyo Trial was not closely examined by academics until the mid-1980s.[51] Richard Minear pointed out in the 1970s that Japanese scholars had been avoiding the Tokyo Trial: 'Where some coverage was unavoidable, they have tended to affirm the validity of the trial and its verdict. Apparently, they fear that denigration of the trial will lead to a positive reevaluation of Japan's wartime policies and leadership.'[52] At the same time, Minear himself acknowledged the pitfall in judging the Tokyo Trial. He emphasized in the Japanese version of his book, *Victors' Justice: The Tokyo War Crimes Trial*, that 'The problem of historical responsibility remains irrelevant to innocence in legal terms.'[53] The difficulty sensed by academics and intellectuals is also shared by the general public. Many interviewees in 2003 said that it was difficult to think and talk about the Tokyo Trial frankly and expressed their hesitation and discomfort in talking about it in public, for fear of being involved in ideological and emotional debates.[54] The majority prefer to distance themselves from the Tokyo Trial and remain silent.

A sense of taboo is therefore an important element, together with ignorance and cynicism, in the national silence over the Tokyo Trial. The silence is vocal in this sense and thus different from 'historical amnesia', as often perceived from outside. At the same time, it can be said that the Japanese were able to remain silent at least until the 1990s, and left a summary of their past war undone for a long time, partly because they were given a judgement on their war by an international tribunal. People could passively stick to the Tokyo Trial's judgement, which was more or less comfortable for them at the time, instead of actively facing and re-examining the painful event by themselves later on. Takashi Ara argued that what was missing there was 'a historical perception from the view of the Japanese people and the pursuit of war responsibility based on it'.[55] According to Noboru Kojima, the 'Tokyo Trial view of history' has put 'a brake on dispassionate research on the history of the period'. Kojima's concern was that the strong influence of the 'Tokyo Trial view of history' 'shrouds the most important aspects of history and leads us to forfeit opportunities for reflection and research'.[56]

The 'history problem' and the Tokyo Trial

The 'history problem', the controversy over Japan's war responsibility, is still a major, sensitive political issue between Japan and China and South Korea.[57] With regard to reconciliation, Yōichi Funabashi pointed out the importance not so much of an apology but of the historical consciousness of why Japan has to apologize.[58] However, the Japanese still hold ambivalent and unsettled view towards the war and have not yet successfully come to terms with their past. This can be seen from the fact that even in the 1990s major Japanese newspapers conducted opinion polls asking the nation questions like 'whether Japan's war was aggression or not' and the result showed a number of people responded 'do not know'.[59]

The 'history problem' was exacerbated at the start of the twenty-first century by the Prime Minister Junichirō Koizumi's annual visit to the Yasukuni Shrine, where the spirits of Tōjō, the 11 convicted by the Tokyo Tribunal and the two defendants that died during the Trial are enshrined together with about 2.5 million war dead since the nineteenth century. Koizumi's visits offended Japan's neighbouring countries from the time he took office in 2001, because it was seen as a total contradiction to express atonement for aggressive war, on the one hand, and to visit the shrine where wartime leaders are enshrined, on the other. In August 2004, the Chinese foreign ministry were thus prompted to release a strong statement: 'The Chinese side hopes the Japanese side will honour its word by *facing up to history*.'[60]

Japanese public opinion regarding Koizumi's visit had been completely divided and the balance between pros and cons fluctuated. *Asahi Shimbun*'s opinion poll in 2001 showed that 41 per cent supported his plan for the first visit, and 42 per cent expressed hesitation in supporting his visit.[61] Opinion became more hostile to the visits when diplomatic relations between Japan and China worsened in spring 2005: 52 per cent felt Koizumi 'should stop his visit', while 36 per cent supported it.[62] On the contrary, in May 2006, 50 per cent supported his visit, while 31 per cent were against.[63] These results, however, do not necessarily reflect the Japanese perception of the Tokyo Trial, nor of the war and war crimes. Although the Tokyo Trial is at the core of the Yasukuni controversy, an opinion poll in May 2006 showed that 70 per cent answered that they 'do not know about the content of the Tokyo Trial'. This implies that to the Japanese the differentiation between war criminals and other war dead in Yasukuni is not necessarily obvious. The Japanese reaction to the Yasukuni controversy well illustrates both ignorance and sensitivity in attitudes towards the past war. It is ironic that the Tokyo Trial, which is regarded as a trial for post-war settlement, has come to be embedded in the 'history problem' and drags Japan back to its dark past.

Conclusion

The nationalist and revisionist critique claims that the Tokyo Trial has implanted in the Japanese psyche a masochistic view of history, which remains deeply ingrained in and harmful to Japanese pride. However, considering that the Tokyo Trial has been not visible in Japanese collective memory and public discourse, it is difficult to trace the active and direct impact of the Trial in cultivating Japanese historical consciousness.

Nevertheless as this chapter has pointed out, the Tokyo Trial had a subtle but negative impact on Japanese historical consciousness. On the one hand, the Tokyo Trial has left fertile ground for emotional debates over the interpretation of the war within a limited circle. On the other hand, the Tokyo Trial contributed, unintentionally, to the majority's 'apathy' towards the war by depriving the Japanese of the will and opportunity to face and judge the country's traumatic past by themselves. What is worse, the Japanese perception of the Tokyo Trial suggests that the Trial's 'victor's justice' has itself become a bitter memory for the nation, which frustrates the nationalists, on the one hand, and creates cynicism and a sense of taboo in many Japanese, on the other. Either way, the Tokyo Trial has been a hindrance to the cultivation of a sound historical consciousness of the past war experience.

In the short run, the Tokyo Trial contributed to the demilitarization and democratization of post-war Japan and achieved its original aim by punishing a handful of wartime leaders. However, its subtle but negative impact on Japanese historical consciousness has led to several problems in the long run. Foremost among them is the way in which Japan has been struggling to achieve reconciliation with its neighbouring countries, who incessantly bring up the 'history problem' and ask for an apology. It seems that not only an ambiguous historical consciousness but also the legacy of the Tokyo Trial itself bear Japan ceaselessly back into the unsettled past.

Notes

An extended version of this chapter has been published in Madoka Futamura, *War Crimes Tribunals and Transitional Justice: The Tokyo Trial and The Nuremberg Legacy* (London: Routledge, 2008).

1. For such an assessment, see, for example, Meirion and Susie Harries, *Sheathing the Sword: The Demilitarisation of Japan* (London: Hamish Hamilton, 1987).
2. This is especially so in the context of the ad hoc International Criminal Tribunals created by the United Nations for the mass atrocities in the former Yugoslavia and Rwanda in the 1990s. See, for example, Antonio Cassese, 'Reflections on International Criminal Prosecution and Punishment of Violations of Humanitarian Law' in J. I. Charney, D. K. Anton, and M. E. O'Connell (eds), *Politics, Values, and Functions: International Law in the 21st Century: Essays in Honor of Professor Louis Henkin*, (The Hague, London: Martinus Nijhoff, 1997); Theodor Meron, 'Answering for War Crimes: Lessons from the Balkans', *Foreign Affairs*, vol. 76, no. 1 (1997), pp. 2–8.
3. Mark J. Osiel, *Mass Atrocity, Collective Memory, and the Law* (New Brunswick, NJ: Transaction Publishers, 1997), p. 6.
4. Takeshi Igarashi and Shinichi Kitaoka (eds), *'Sōron' Tōkyō Saiban towa Nandattanoka* (Tokyo: Tsukijishokan, 1997), p. v.
5. For an overview of the Tokyo Trial and its procedures, see, for example, Richard H. Minear, *Victors' Justice: The Tokyo War Crimes Trial* (Princeton, NJ: Princeton University Press, 1971); Chihiro Hosoya, Nisuke Andō, Yasuaki Ōnuma and Richard Minear (eds), *The Tokyo War Crimes Trial: An International Symposium* (Tokyo: Kodansha; New York, NY: Distributed in the US by Kodansha International through Harper & Row, 1986); Neil Boister and Robert Cryer, *The Tokyo International Military Tribunal: A Reappraisal* (Oxford: Oxford University Press, 2008). The author learned a lot from Yuma Totani's

The Tokyo War Crimes Trial: The Pursuit of Justice in the Wake of World War II (Cambridge, MT and London: Harvard University Asia Center, 2008), which, unfortunately, was published after this chapter, as well as the author's monograph, was written. I am grateful to Dr Totani's comment on my monograph, pointing out arguable aspects of the received fact of the Tokyo Trial, which has been widely taken for granted by researchers. In the following footnotes, I note these points, although in a limited form.

6. When it comes to the actual prosecutorial procedures, prosecutors from 11 countries all played a role. See Totani, *Tokyo*, Chapter 7.
7. Politico-Military Problems in the Far East: United States Initial Post-defeat Policy Relating to Japan (SWNCC150/4/A), 21 September 1945.
8. Quoted in Minear, *Victors'*, p. 126, Footnote 3, emphasis added.
9. Harries, *Sheathing*, p. xxix.
10. 'The Judgment' in R. John Pritchard and Sonia M. Zaide (eds) and Project Director: Donald C. Watt, *The Tokyo War Crimes Trial: The Complete Transcripts of the Proceedings of the International Military Tribunal for the Far East in Twenty-two Volumes* (New York and London: Garland Publishing Inc., 1981) (as *Transcripts* hereafter), Vol. 20, pp. 49,765.
11. Minear, *Victors'*, p. 134.
12. 'Summation by the Prosecution', 11 February 1948, *Transcripts*, Vol. 16, pp. 38,972.
13. Makoto Iokibe, '"Tōkyō Saiban" ga Sabaita Hito to Jidai' in Kōdansha (ed.), *Tōkyō Saiban: Shashin Hiroku* (Tokyo: Kōdansha, 1983), p. 109.
14. Quoted in Donald Cameron Watt, 'Historical Introduction' in R. John Pritchard and Sonia M. Zaide (eds) and Project Director: Donald C. Watt, *The Tokyo War Crimes Trial: The Comprehensive Index and Guide to the Proceedings of the International Military Tribunal for the Far East in Five Volumes* (New York and London: Garland Publishing Inc., 1987), Vol. I, p. xviii.
15. Iokibe, '"Tōkyō"', p. 111. Konoe committed suicide before arrest. Matsuoka was one of the 28 defendants of the Tokyo Tribunal but died immediately after the opening of the Trial. Iokibe also pointed out the absence of Kanji Ishihara, a military officer and the mastermind of the Manchurian Incident. He argues that the absence of these three figures degraded the standard and accuracy of the Tribunal's historical record. Ibid., p. 108.
16. Quoted in Masanori Nakamura, 'Shōwashi Kenkyū to Tōkyō Saiban' in Igarashi and Kitaoka (eds), *'Sōron'*, pp. 181–2.
17. See Telegram, General of the Army Douglas MacArthur to the Chief of Staff, United States Army (Eisenhower), 25 January 1946 in *F.R.U.S.*, 1946, Vol. VIII, p. 396. Unlike many of existing studies argue, Totani points out that MacArthur, in spite of this often-cited telegram, had no power to make decisions on the trials of the Emperor. See Totani, *Tokyo*, Chapter 2.
18. Kentarō Awaya, *Tōkyō Saibanron* (Tokyo: Ōtsuki Shoten, 1989), pp. 269–70. Based on this view, there has been a widespread understanding that the Tribunal failed to examine many of the alleged Japanese war crimes conducted against Asian civilians under Japanese colonial rule. On the contrary, Totani's latest research, through a thorough examination of the Trial transcripts, pointed out that the Allied prosecutors did try to substantiate Japanese alleged war crimes against Asian civilians, including sexual slavery, known as 'comfort women'. However, these efforts were not reflected in the

final judgement, and Japanese leaders were not held accountable for such crimes. See Totani, *Tokyo*, Chapter 7.

19. Quoted in Yutaka Yoshida, 'Senryōki no Sensō Sekininron' in Ajia Minshūhōtei Junbikai (ed.), *Toinaosu Tōkyō Saiban* (Tokyo: Ryokufū Shuppan, 1995), p. 248.
20. Preface in Hosoya, Andō, Ōnuma and Minear (eds), *Tokyo*, p. 8, emphasis added.
21. R. John Pritchard, *An Overview of the Historical Importance of the Tokyo War Trial*, Nissan Occasional Paper Series No. 5 (Oxford: Nissan Institute of Japanese Studies, 1987), p. 50.
22. Minear, *Victors'*, p. 159; Watt, 'Historical', p. xxii.
23. William Joseph Sebald, *With MacArthur in Japan: A Personal History of the Occupation* (London: Cresset Press, 1967), p. 151.
24. Quoted in Yoshiaki Yoshimi, 'Senryōki Nihon no Minshū-Ishiki: Sensō-Sekininron wo Megutte', *Shisō*, No. 811 (January 1992), 77.
25. Shunsuke Tsurumi, *A Cultural History of Postwar Japan: 1945–1980* (London and New York: KPI Limited, 1987), p. 15.
26. The serial was published in newspapers, such as *Asahi Shimbun* and *Yomiuri Shimbun*, during 8–17 December 1945.
27. According to the opinion poll conducted by *Yomiuri Shimbun* in August 1948, 68.5 per cent supported the Emperor remaining in his status. Quoted in Tōkyō Saiban Handobukku Henshū Iinkai (ed.), *Tōkyō Saiban Handobukku* (Tokyo: Aoki Shoten, 1989), p. 233. The general support, Kiyoko Takeda points out, was based on the fact that people were 'favourable to the person of the emperor and not to an absolute and authoritarian emperor system'. Kiyoko Takeda, *The Dual-Image of the Japanese Emperor* (Basingstoke: Macmillan, 1988), p. 150.
28. Quoted in John W. Dower, *Embracing Defeat: Japan in the Aftermath of World War II* (London: Penguin Books, 2000), p. 490.
29. See Carol Gluck, 'The Past in the Present' in Andrew Gordon (ed.), *Postwar Japan as History* (Berkeley: University of California Press, 1993), pp. 66–70.
30. For example, a vocal anti-Tokyo Trial critic Masaaki Tanaka wrote a book, *Nankin Gyakusatsu no Kyokō* (The Fiction of the 'Nanjing Massacre') (Tokyo: Nihon Kyōbunsha, 1984).
31. Resolution to Renew the Determination for Peace on the Basis of Lessons Learned from History, House of Representatives, National Diet of Japan, 9 June 1995: http://www.mofa.go.jp/announce/press/pm/murayama/address9506.html, 15 October 2004. See also Murayama's more unequivocal apology: http://www.mofa.go.jp/announce/press/pm/murayama/9508.html, 15 October 2004.
32. Tōru Maeno, *Sengo Rekishi no Shinjitsu* (Tokyo: Keizaikai, 2000), p. 23.
33. For details of the movement, see Gavan McCormack, 'The Japanese Movement to "Correct" History' in Laura Hein and Mark Selden (eds), *Censoring History: Citizenship and Memory in Japan, Germany, and the United States* (Armonk, NY: M.E. Sharpe, 2000), pp. 53–73.
34. The association's textbook became the source of diplomatic tension again in 2005, when its latest version passed the checking and certification process of the Ministry of Education, Culture, Sports, Science and Technology.
35. Quoted in Nakamura, 'Shōwashi', p. 184.
36. There is a slight generation gap in perception of the past war, which, up to a point, was supported by people at the time. See footnote 38 below.

37. For the 'contested memories' of the Japanese over the war and war crimes, see John W. Dower, '"An Aptitude for Being Unloved": War and Memory in Japan' in Omer Bartov, Atina Grossmann, and Mary Nolan (eds), *Crimes of War: Guilt and Denial in the Twentieth Century* (New York: The New Press, 2002), pp. 217–41.
38. *Mainichi Shimbun*, 15 August 2005. According to an opinion poll conducted by *Yomiuri Shimbun* in October 1993, 53.1 per cent thought 'Japan was an "aggressor" in WWII', while 24.8 per cent did not agree with this view. Importantly, in every generation from those in their 20s to those over 70, those who agreed with the statement exceeded those who disagreed, although the percentage of those who agree to the statement is highest among those in their 20s (61.7 per cent), and lowest among those over 70 (41.1 per cent). *Yomiuri Shimbun*, 5 October 1993, quoted in ibid., p. 315, footnote 5.
39. *Asahi Shimbun*, 23 August 1994.
40. *Asahi Shimbun*, 2 May 2006.
41. Twenty-one intensive interviews with individuals ranging from their 20s to late 60s and five focus groups were conducted originally for the author's Ph.D. thesis: Madoka Futamura, *Revisiting the 'Nuremberg Legacy': Societal Transformation and the Strategic Success of International War Crimes Tribunals – Lessons from the Tokyo Trial and Japanese Experience* (King's College London, 2006).
42. Daizaburō Yui, 'Komento' in Igarashi and Kitaoka (eds), *'Sōron'*, p. 44.
43. Yasuaki Ōnuma, *Tōkyō Saiban kara Sengo Sekinin no Shisō he*, 4th edn (Tokyo: Tōshindō, 1997), p. 11.
44. Yui, 'Komento', p. 44.
45. See Gluck, 'Past'.
46. Barrie Paskins and Michael Dockrill, *The Ethics of War* (London: Duckworth, 1979), p. 266.
47. *Asahi Shimbun*, 8 February 1989.
48. For the Japanese sense of individual and collective guilt, see Futamura, *War*, Chapter 6.
49. The editorial of *Mainichi Shimbun* raised this point on 23 December 1948, the day seven defendants were executed.
50. Aiko Utsumi in Ajia Minshūhōtei Junbikai (ed.), *Toinaosu*, p. 3; Nakamura, 'Shōwashi', p. 185. Unlike the 'Second World War' and the 'Asia-Pacific War', the 'Fifteen-years war' and the 'Greater East Asia War' are loaded terms, which carry the political and ideological stance of those who utilize them: the former emphasizes the aspect of Japan as an aggressor towards its Asian neighbours, while the latter connotes the justification and glorification of the war as self-defence and the liberation of Asia.
51. According to the National Diet Library's database, 174 books, appeared under key words either '*Tōkyō Saiban* [the Tokyo Trial]' or '*Kokusai Gunji Saiban* [the International Military Tribunal]' excluding Nuremberg, were published between 1946 and 2003. Among them 23 books, including court materials, were published during the occupation (1946–51); 13 during the 1950s; 14 in the 1960s; and 21 in the 1970s. The number suddenly went up in the 1980s, during which 36 were published, out of which 33 were published after 1983, the year a feature-length documentary film of the Tokyo Trial was released and achieved great success. In 1990s, there were 54 publications, and after

2001 up until 2003, there were 13. Nearly 60 per cent of the publications on the Tokyo Trial up until 2003 were published after 1983.

52. Minear, *Victors'*, p. ix.
53. Richard H. Minear, translated by Nisuke Andō, *Shōsha no Sabaki: Sensō Saiban, Sensō Sekinin towa Nanika* (Tokyo: Fukumura Shuppan, 1972), pp. 5–6. The difficulty of criticizing the Tokyo Trial constructively can also be seen from the fact that Judge Pal at the Tribunal, who wrote a dissenting opinion challenging the jurisdiction of the Tokyo Tribunal and acquitting every defendant, has been a favourite figure for the nationalists and revisionists.
54. See Futamura, *War*, Chapters 5 and 6.
55. Takashi Ara, 'Tōkyō Saiban: Sensō Sekininron no Genryū – Tōkyō Saiban to Senryōka no Yoron', *Rekishi Hyōron*, No. 408 (April, 1984), pp. 19–20.
56. Noboru Kojima in Hosoya [et al.] (eds), *Tokyo*, pp. 77–8.
57. See Yoichi Funabashi (ed.), *Reconciliation in the Asia-Pacific* (Washington, DC: United States Institute of Peace Press, 2003).
58. Yōichi Funabashi in '"Sensō Sekinin" no Chakuchiten wo Motomete', *Chūōkōron* (February 2003), p. 69.
59. *Yomiuri Shimbun*, 5 October 1993.
60. 'Japan shrine visit angers China', BBC news, 15 August 2004, http://news.bbc.co.uk/go/pr/fr/-/1/hi/world/asia-pacific/3567084.stm, emphasis added.
61. *Asahi.com*, http://www.asahi.com/special/shijiritsu/TKY200404190343.html, 1 July 2005.
62. *Asahi.com*, http://www.asahi.com/special/050410/TKY200506270317.html, 1 July 2005.
63. *Asahi Shimbun*, 2 May 2006.

10

Japanese War Veterans and *Kamikaze* Memorialization: A Case Study of Defeat Remembrance as Revitalization Movement

M. G. Sheftall

This chapter of *Defeat and Memory* will examine aspects of Asia-Pacific War remembrance in modern-era Japan – a bitterly contested discourse of historical interpretation well into its third generation and still far from resolution. Specifically, the focus will be on the participation of two Japanese war veterans' or war veteran-dominated organizations in the ongoing and still contested historical discourse of Japan's wartime *kamikaze* programme, which can serve as a useful microcosm for the discursive tone and key issues of Japanese defeat remembrance as a whole.

Historical context

Before we examine the actors and issues involved in *kamikaze* remembrance, it may behove us to pause here to consider the historical narrative in question. Although the meaning or interpretation of this narrative is contested, the following basic facts are generally acknowledged (albeit with considerable variation in wording) by all participants in the discourse:

- From October 1944, the Imperial Japanese military – choosing to continue fighting in a combat environment in which air superiority had been completely and irretrievably ceded to the enemy – began fielding what were, in dispassionate terms of tactical effect, guided missiles. The technology that could be applied to these 'guided missiles', however, was limited to what was available to the Japanese military in 1944–5.
- Technological limitations, the exigencies of a rapidly deteriorating tactical situation and, perhaps, the temptation of having a ready supply of fighting men on hand that were capable of any sacrifice for

their cause led Japanese planners to the conclusion that a human brain and body – that is, a pilot – comprised the most pragmatic and effective option available for the target acquisition and guidance control system necessary for the new weapon, while a bomb-laden high performance military aircraft was the best choice for its ballistic component. 'Successful' deployment of the weapon, of necessity, involved the inevitable destruction of the human 'guidance system' when the 'missile' crashed into its target – generally an Allied warship.

- There was, initially, a measure of opposition within the military to the expending of human lives in such a horrific manner. However, the success of the tactic's debut during the Battle of Leyte Gulf in October 1944 – and perhaps of equal importance, the lockstep conformity that characterized Japanese military culture of the era – silenced most critics. From then on until the end of the war it was acknowledged by the overwhelming majority of field commanders in both the Imperial Navy and Army that the new 'missile' was capable of accuracy and destructive power greater than what was by then considered obtainable with conventional air tactics against superior Allied forces.
- Both the tactic and the personnel tasked with employing it in combat are known to history as the *kamikaze* or 'divine wind', but to Japanese they were and generally still are referred to as *tokkō*, an abbreviation of a wartime euphemism for self-immolatory tactics that translates simply as 'special attack'. For the most part, young, unmarried pilots with little or no previous combat experience carried out the *kamikaze* missions. Ostensibly 'volunteers', these young men were told over and over again not only by their military superiors but also by the mass media apparatus at the disposal of the state that – outside of the extreme unlikelihood of ultimate victory against the Allies – there could be no greater glory for Japanese fighting men than to sacrifice their lives in suicide attacks to defend the sacred soil of their homeland as a proud nation prepared to choose 'the honourable death of 100 million' – the ominously-if-poetically-termed *ichioku gyokusai* – over the ignominy of surrender and foreign occupation. By war's end, some 5843[1] of these young men had died in *kamikaze* attacks that sank or damaged over 200 Allied ships and killed or wounded approximately 15,000 Allied servicemen.[2]

Framing the discourse of *kamikaze* veteran remembrance

The discourse under examination in this study is limned by the respective (but nevertheless extensively overlapping) 'preferred readings' of the narrative outlined above, as well as by the organizational structures, pedagogical agendas and ritual activities of two of the most historically important and influential agents in *kamikaze* memorialization.

The Special Attack Memorial Association (*Tokkō Zaidan*, hereafter TZ) has a distinctly 'revitalizationist' agenda and orientation.[3] It is composed of veterans of wartime *kamikaze* units, bereaved family members of *kamikaze* pilots killed in action and a third 'interested party' subgroup, the members of which have no personal connection to the wartime *kamikaze* programme (and who in many cases have no memory of – or were not even born until after – the war).

The other group under examination is the Divine Thunder Unit Veterans Association (*Jinrai Butai Senyūkai*, hereafter JBS). The organizational tone of this group might be best described as 'historical realist'. Its membership is comprised solely of veterans of an elite Imperial Navy unit that deployed piloted Ōka rocket-boosted glider bombs in the closing months of the war.

Despite significant differences in terms of respective agenda and structure, both groups can be considered 'revisionists' (although the term applies comparatively loosely to the JBS), defined for purposes of this chapter as 'advocates of a . . . [more] positive interpretation of Japan's cause and conduct in the Asia-Pacific War' than is currently accepted as the norm in mainstream Japanese public opinion.[4] The TZ's revisionist efforts are carried out within the framework of the group's traditionally religious orientation, putting heavy emphasis on ritual and propounding a Buddhist/Shintō theological justification for the *kamikaze* within an overarching context of the war as having been fought by Japan on Asia's behalf against the encroachment of Western imperialism – an interpretation of war aims identical with that espoused in wartime Japanese propaganda. An encapsulation of the TZ's 'preferred reading' of the *kamikaze* narrative in the overall context of the war/defeat might be, 'Japan actually won the war by losing it so honourably, and in a manner of which only Japanese are capable'. Additionally, in its religious ceremonies, the group propounds the theological concept – shared by Yasukuni Shrine – that upon their deaths in battle, the *kamikaze* pilots underwent an apotheosis to become nation-protecting deities who are owed the eternal gratitude of the Japanese people not only for the pilots' noble sacrifices but also for making possible Japan's post-war peace and prosperity through divine intervention.[5]

The main pedagogical goal of the TZ, manifest in the highly political content of the group's publications and prayers as well as in speeches by both members and invited VIPs delivered at its various group events, is the dissemination among younger generations of Japanese of this 'preferred reading' – a stance that puts the group in close ideological step with the official position of Yasukuni Shrine and other agents clustered on the right end of Japan's defeat remembrance discursive continuum. In addition to seeking wider public sympathy towards a historical rehabilitation of Japan's war aims and effort, the TZ also sees (and thus also seeks to portray) the years of the Asia-Pacific War as a halcyon era of Japaneseness (especially Japanese masculinity), the ways and mores of which should be revived as an antidote to what the majority of the group's members (and a rock solid consensus of its officers) see as the decadent and perilous condition of nihilistic ennui, vapid materialism and rampant individualism currently plaguing post-war Japanese society.[6]

The 'realist' JBS, on the other hand, is comprised entirely of veterans and thus constitutes a 'veterans' group' in the more traditional sense of the term. In addition to formal memorialization, its main activities are maintaining personal ties; collecting money for floral displays and so on, for members' funerals; updating the unit's wartime service record; and cooperating with historical researchers, both Japanese and non-Japanese. On a more personal level for the JBS members, it is not difficult to perceive that a desire to celebrate the memory of their shared camaraderie during the war is also an important reason for their membership in the group.

Compared with the activities of the revitalizationist TZ, the JBS puts far less emphasis on ritual and almost none at all on theological matters outside of the requirements of prayers offered for dead comrades at group gatherings. The members of the JBS are also less interested in rationalizing the war or the *kamikaze* concept than they are in honouring the memory of dead comrades as individuals and ensuring that an accurate historical account of these comrades' feats of arms is left for posterity.[7] Accordingly, the group's 'preferred reading' of the *kamikaze* narrative could be summed up thus: 'As witnesses and survivors, we must never let something like the *kamikaze* happen in Japan again – but neither must we allow the bravery and sacrifices of our comrades to be forgotten.' If pressed, most of the JBS members would probably also acknowledge some aspect of Japanese cultural exclusivity – and thus some grounds for ethnic pride, however shaky – in the *kamikaze* phenomenon, although perhaps without the same ardour and exuberance with which TZ members express this sentiment.[8]

The nature of personal relations and connections within the two groups is also quite different. While the JBS members have all known each other for over 60 years, the TZ's members' association with one another, with the exception of some of the veterans (particularly among service academy class year groups), is fundamentally task oriented and dates mostly from the post-war era – in many cases, only from the 1980s, 1990s or even later. This membership flexibility and constant influx of new members, in addition to its hierarchical structure and its well-coordinated pedagogical efforts, afford the TZ aspects one might associate more with a political organization than with a 'memorial' group.

These and other differences aside, the respective efforts of the groups can be placed into two distinct though overlapping categories of interest:

1. both groups are intimately concerned with *kamikaze* memorialization;
2. both have an agenda of promoting wider public acceptance of a *kamikaze* narrative interpretation that is both iconic of, and inclusive to, the overarching context of improving the image of Japan's cause and conduct in the Asia-Pacific War (but in the case of the JBS, the iconography currently being crafted is by no means exclusively or unabashedly hagiographic).

Kamikaze remembrance as revitalization rite: Old heroes for new problems

So, then, to what halcyon past do Japanese revisionists and *kamikaze* hagiographers seek to return? While only the most unhinged nostalgic apologist for twentieth- century Japanese militarism would yearn for the nation's citizens to be given a chance to prove their mettle and loyalty in another war, the notion that, if a need arises, the nation should have a population base of young men at its disposal capable of making such sacrifices – much as their grandfathers and great-grandfathers did in the Second World War – has not insignificant appeal, especially among 30-and-under Japanese males.[9] At the very least, the *kamikaze* pilots can be held up as models of ethnically exclusive masculine virtue in an overarching revisionist historical context, forming a crystallization point around which an alchemy for ego-protecting and emotionally gratifying Asia-Pacific War defeat remembrance can coalesce. Used to such ends, the revisionist/hagiographer can utilize *kamikaze* iconography to

1. emphasize Japanese martial prowess, but in a defensive rather than a more morally questionable offensive context;
2 repair damage to Japanese (and particularly male) pride suffered from the defeat and subsequent Allied occupation;

3. emphasize Japan's war with the West – especially with superpower America depicted as a bastion of Anglo-Saxon global hegemony – obfuscating or otherwise minimizing (at least in popular Japanese historical consciousness) Japanese victimization of other Asians;
4. espouse an ego-gratifying 'we only lost to overwhelming numbers' defeat rationalization;
5. cast the war in a populist 'by the people, for the people' light by showcasing the deeds of the Japanese rank-and-file citizen-soldier rather than the activities of high level policymakers (imagery that can lay blame on a political elite for Japan's calamity – payment in full for which is symbolized by the blood sacrifice of the *kamikaze* – while simultaneously alleviating the state's burden of responsibility by spreading duplicity for the *kamikaze* concept as widely and evenly as possible in referring to the pilots as 'volunteers');
6. arouse sympathy in generations of Japanese unfamiliar with the war, but more in a sense of gratitude and admiration for the sacrifices of its fighting men than of pity over their fate; and
7. portray the *kamikaze* as embodying the finest aspects of an idealized national character (*kokuminsei*), particularly loyalty, bravery, purity of spirit and commitment to selfless sacrifice (*messhi hōkō*), thereby evoking a comforting communal sense of historical continuity with a halcyon Japanese past.

But even if groups bent on singing paeans to the *kamikaze* could secure a public platform for such appeals, the non-Japanese reader may have difficulty fathoming how these views could possibly find a sympathetic audience in a purportedly peace-loving, democratic and highly economically developed twenty-first-century society ostensibly so anti-military – at least on the paper on which its Constitution is written – that it cannot even legally refer to its own armed forces by name as such.

That *kamikaze* encomium can indeed find such an audience – and a ready one, at that – is likely due to several interrelated factors. The two most important of these are

1. cultural empathy and pride (which has a tendency to flow over into ethnocentric excess, as constantly demonstrated in Japanese popular culture portrayals of *kamikaze* thematic material)
2. and collective anxiety over a perceived decline in national/cultural prestige never quite shaken off since the defeat of 1945, and coming back with a vengeance since the economic downturn of the early 1990s.

While the wartime exploits of the *kamikaze* tinged Western and especially American stereotypes of Japanese 'Otherness' with a fanatic or even lunatic aspect that shows little sign of fading even 60 years after the events of 1944–5, the *kamikaze* legacy has developed quite differently in Japan's post-war collective memory. Whether the message is ideological and somewhat abstract, as when the *kamikaze* are utilized as coded signifiers of culturally exclusive Japanese spiritual superiority – a common theme for revisionism of the TZ/Yasukuni Shrine variety – or more along the lines of the personalized portraiture of remembered and beloved comrades-in-arms that tends to be favoured in the memorialization activities of veterans' groups like the JBS, the message is capable of finding a sympathetic audience among a strikingly diverse cross-section of Japanese. This appeal is no better evidenced than by the hundreds (or perhaps even thousands) of Japanese novels, TV dramas, *manga* comic books, cartoons and popular films that have appeared in the post-war in which the *kamikaze* narrative serves either as the central theme or as a dramatic plot hook of the work.

The concept of irrational yet heroic and morally blameless victimization – an archetypal trope in Japanese lore for well over a thousand years, according to Ivan Morris – may be useful in understanding the soft spot in the Japanese heart for the legacy of the *kamikaze*.[10] For the better part of the last six decades, the tragic-yet-resolute figure of the *kamikaze* pilot has stood front-and-centre with the atomic bombings of Hiroshima and Nagasaki, the humiliation of foreign occupation and the privations of post-war life to hold a position of honour in the sepia-toned pantheon of Japanese World War II victimhood iconography. Concurrently, his fiery sacrifice has stood as potent symbolism both for the self-immolatory denouement of Japan's disastrous experiment in militaristic ultranationalism, and in a more romantic sense, to support the notion that some primal, indivisible 'Japanese spirit' has survived unscathed from the ages of the ancient Yamato clan and the days of the samurai into the modern era.

However, this culturally reaffirming, politically 'safe' and (therefore, by current Japanese definition) distinctly anti-militarist post-war trope of '*kamikaze* pilot as victim' – virtually canonical at least since the publication and subsequent explosive popularity of the anthology *Kike Wadatsumi no Koe* in 1949[11] – has during the last 15 years or so begun to give ground to more laudatory portrayals of the *kamikaze* pilot as 'hero' or even 'role model' that seek to exonerate, in the same breath, not only the motives of Japan's wartime leadership but also the nation's complicity in waging the war itself.[12] It is clear that this movement towards a more hagiographic appraisal of the historical legacy of the *kamikaze* is one facet of a distinctly perceptible increase in popular receptivity during this period towards more positive interpretations of Japan's role in the Asia-Pacific War.

It is tempting to see these trends since the early 1990s as the nascent stirrings of what could be considered a *revitalization movement*, defined by anthropologist Anthony F. C. Wallace as a 'conscious, deliberate, organized effort on the part of some members of society to create a more satisfying culture', returning to 'a golden age . . . a previous condition of social virtue',[13] with the longed-after Japanese Camelot in this case being a state and society that could boast (as it once did) of millions of young men willing (if not eager) to die for their country. Moreover, attempts to lionize the *kamikaze* and enlist this revamped legacy in a populist/nativist campaign to instil patriotic/ethnic pride in younger generations of Japanese fit the model of 'revivalistic nativistic movements' defined in anthropologist Ralph Linton's classic study as 'associated with frustrating situations [that] are primarily attempts to compensate for the frustrations of the society's members. The elements revived become symbols of a period when the society was . . . in retrospect, happy or great'.[14]

As Yoshida Yutaka and John Nathan point out, the loss of national confidence the Japanese people have experienced during the 15-year-long economic slump following the collapse of the 'Bubble Economy' in 1990 would seem to indicate that the timing of the revisionist movement is not coincidental.[15] Moreover, in the interim, conditions favourable to Japanese revisionism have been exacerbated by anxiety at all levels of society and government over the end of the Cold War Asian balance of power, the subsequent meteoric rise of up-and-coming economic and military superpower China (which still bears grudges from 1931–45 and is not above utilizing them both politically and diplomatically) and bitterly self-conscious impotence vis-à-vis regular North Korean military and diplomatic provocation. But then again, it would be a mistake to interpret this revisionism as a sign that the Japanese are psychologically preparing themselves to start gassing up their *kamikaze* planes again. Rather, it seems clear that this general but gradual rightward shift in Japanese political sentiment and historical interpretation is not so much an empowerment fantasy as it is a collective psychological defence mechanism against anxieties over the current and future status of Japan and its culture.

Seen within this post-war historical and socio-economic context, it may be closer to the crux of the matter to regard the popular fascination of late with the iconic figure of the *kamikaze* pilot as being less a matter of specifically longing for Japanese heroes from 1944–5 than of being unable to find any Japanese heroes from the post-war era that are capable of calming the fears and combating the malaise many Japanese see afflicting their society on individual and collective levels. What is sought after in this anxious nostalgia – and reality-distancing nostalgia is what this or any other revitalization movement really boils down to in the end, all being said – is not the frisson of desperate action, but rather, the comforting (if, to an extent, melancholy) lull of a culturally reaffirming dream. In the next two sections, this chapter will trace the origins of this twenty-first-century dream to a midsummer afternoon's nightmare in 1945.

Post-war ruins, *Kyodatsu* and the emergence of the JBS

While it is common knowledge that the *kamikaze* caused great devastation at the cost of thousands of human lives on both sides of the conflict, it is not so well known – at least outside of Japan – that additional thousands of fully trained, well-equipped and thoroughly indoctrinated *kamikaze* personnel – aircraft pilots, rocket bomb and human torpedo crewmen and suicide frogmen – were still alive when Emperor Hirohito ordered his armed forces to lay down their arms to end the war. As His Majesty's surrender broadcast faded out into crackling static on radio sets across the nation shortly after noon on 15 August 1945, thousands of suddenly unemployed and de-apotheosized young war gods were now relegated to the humbled earthbound status of mere mortals facing, along with millions of exhausted, grieving, disillusioned countrymen, the thankless job of clearing the scorched rubble of defeat to begin rebuilding their shattered homeland.

But the ruins the *kamikaze* programme's survivors found upon their return home were not just physical. What also lay in ruins in the aftermath of the surrender broadcast were the *kamikaze* survivors' identities as elite warriors. Moreover, in a sense shared with the six million other Japanese soldiers, sailors and airmen in uniform (or some tattered semblance of same) now wending their way home, their sense of self-worth as men, or even as human beings, had also taken a terrific beating from their collective experience of war and defeat.

The 'welcome' most *kamikaze* veterans received from their civilian compatriots upon their return home only compounded this angst. In the aftermath of surrender, the change in Japanese public opinion towards the *kamikaze* programme and anyone and anything associated with it was swift and merciless. Up until the last days of the war (and in fact, until the very day of surrender) the *kamikaze* personnel – lauded in every mass media organ at the disposal of the state as 'divine eagles' (*kami-washi*) and 'sacred shields of the Imperial realm' (*kōkoku no mitate*) – were held up as paragons of idealized masculine virtue and samurai spirit.[16] Scant weeks later, as the same mass media organs – now under GHQ/SCAP[17] supervision – made the phrase '*kamikaze* breakdown' (*tokkō kuzure*) household words across the nation; the *kamikaze* survivors were viewed increasingly and alternately as ciphers, wraiths or walking time bombs. In an editorial in the influential daily *Asahi Shimbun* on 16 December 1945, one of the greatest living figures in Japanese letters at the time, Shiga Naoya, wrote,

> It is the height of irresponsibility for the government to demobilize and return to society in their current mindset the young former *kamikaze* personnel who have undergone special psychological indoctrination . . . [I]t chills the heart to contemplate the deeds of which they may be capable . . . The potential for societal disruption must be considered.[18]

Other commentators – less erudite but perhaps more sympathetic than Shiga – nevertheless painted a portrait of '*kamikaze* breakdown' that was no less pitiful. In the *Asahi Shimbun* 'Voice' (*Koe*) column of 13 January 1946, which featured letters from readers about the despondence of *kamikaze* survivors, a Tokyo high school girl wrote,

> The *kamikaze* pilots – called 'divine eagles' not so long ago – are coming home after a well-fought struggle. And some of them have been arrested recently after committing terrible crimes. However, our hearts cannot help but ache when we read accounts such as the one about a recently arrested and utterly despondent 17-year-old veteran, who explained his duplicity in a crime as having been the result of being misled under the influence of a 'bad friend'. After a long hard fight on foreign soil, veterans are returning home to the cold stares of their neighbours. Should not the people of the nation reflect on their share of blame in their treatment of the returning veterans as losers?[19]

In the same column, a recently demobilized student draftee from Akita Prefecture wrote,

> I could not contemplate without tears the awful turn the world has taken when I read a recent newspaper article reporting on the arrest

> of a former *kamikaze* pilot on charges of robbery. Just yesterday the *kamikaze* pilots were regarded as living gods, but now they have fallen to such pitiable ignominy.[20]

In his essay 'On Decadence', Sakaguchi Ango, perhaps, most succinctly captured the spiritual oblivion and sheer, crushing anticlimax ofthe *kamikaze* survivor's post-war reality – and the disillusionment other Japanese experienced when contemplating his fate – when he wrote, 'Could we not say that the *kamikaze* hero was a mere illusion, and that human history begins from the point where he takes toblack-marketeering?'[21]

For Suzuki Hideo, *kamikaze* survivor and (unofficial) chairman of the JBS since its founding in 1951, the dominant memory of the first few months of his post-war 'second life' is *kyodatsu* – the bottomless despair and disillusionment experienced by almost all Japanese in the post-war after coming down off the powerful high of fatalistic late-war ultranationalism.[22] After his homecoming, Suzuki spent the rest of 1945 and several months into 1946 in near utter seclusion in his old bedroom, variably sleeping, reading and afflicted by sudden crying jags. 'Almost every man in the country suffered from *kyodatsu*', Suzuki recalled in a March 2006 interview with the writer. 'Shrewder types of fellows [*yōryō no ii hito*] quickly bounced back onto their feet, getting involved in the black market and all, but I was not one of them.'

One hope that kept Suzuki going through his first few post-war years was a promise he and the other members of the Divine Thunder Unit had made on the day of their demobilization, 23 August 1945. The pledge had been to meet at Yasukuni Shrine – if the facility still existed by then – on 21 March 1948, the third anniversary of the unit's disastrous combat debut during the opening stage of the Okinawa Campaign. The reunion went as planned on the appointed day, with prayers at Yasukuni followed by an overnight stay at a hot springs hotel rented out under the innocuous group name of *Hagoromo Kai* (a pseudonym being necessary, as formal veterans' gatherings were frowned upon by GHQ/SCAP authorities). The group – similar to the practice of other post-war veterans' organizations – only began using a name evocative of its members' wartime unit affiliation after the end of the Occupation made such public use of wartime terminology possible.[23]

At the conclusion of the March 1948 reunion, the attendees promised to repeat the event after another three-year interval, meeting again at Yasukuni on 21 March 1951. Suzuki and a handful of the other non-academy graduate officers naturally gravitated towards leadership roles.[24] Several men who had been wartime clerks with the Jinrai unit became the administrative staff of the group, their primary tasks during the subsequent three-year interval and thereafter being to maintain a member roster and mailing address list, facilitate communication between members and bereaved relatives of comrades and track down other unit members whose status and whereabouts were as yet unknown.

One of these ex-clerks, Torii Tatsuya, also the hereditary proprietor of a printing firm, and his efforts proved especially valuable towards the publication of a unit history/remembrance volume in 1952. Written a scant six years after the war, the letters compiled in the volume overflow with emotions that are palpably and understandably raw – some writers are proud, almost boastful, of military prowess, more are morose and contrite over the defeat, or even over their own survival. More remarkable, however, is the candour of some of the letters, which are breathtakingly frank in their criticism of the *kamikaze* concept specifically and of the war in general. With the distance of time, the fossilization of taboos and group values and self-conscious restraint on the part of the veterans out of consideration for the feelings of bereaved family members, the expression of such sentiments and in such a tone has long since been rendered well nigh impossible. For this reason, as well as the fact that the book is the only compilation of its kind published by *kamikaze* veterans during the Occupation, the historical value of this document is unquestionably high.

After the 1951 gathering, the members began meeting annually, and as their membership roster grew, so too did the scale and ambition of their memorial activities. One of the earlier and more unusual organized events of the newly renamed *Jinrai Butai Senyūkai* was a 'flight experience' (*hikō taiken*) session held in 1954 for bereaved family members of unit personnel killed in the war. At the time, very few Japanese without military service had ever flown in an aeroplane, and it was felt that the experience of flight would provide closure for family members who previously could only imagine what their lost loved ones (sons, husbands, brothers) had experienced during their time as naval aviators. For the occasion, the JBS chartered a small single-engine Cessna plane flying out of Shinonome Airfield in Tokyo to take groups of four or five people at a time on short flights over the capital.

Over the years, as the war-weary young men of 1945 approached middle age and enjoyed a commensurate increase in professional success and social influence, the group's standing with Yasukuni grew as well. In 1957, the JBS was granted permission to plant four cherry trees on the grounds of the shrine. The trees are still there, located on the right side of the main gate when approaching the main altar, and one of them – a particularly florid *someiyoshino* cherry – is the Japan Meteorological Agency's official marker tree for its annual announcement of the beginning of cherry blossom season in Tokyo.[25]

Suzuki believes the JBS reached its peak in membership and in 'esprit de corps' probably around the time of the 21 March 1977 gathering, on the occasion of which the association voted to raise money for the commissioning of a full-scale *Ōka* glider bomb replica model and a miniature scale panoramic diorama depicting the first *Ōka* mission. Both items were donated to Yasukuni Shrine upon completion. The unveiling ceremony for these items was held at Yasukuni on 21 March 1979, and the visit to the exhibition a month later by Prince and Princess Takamatsunomiya and Princess Chichibunomiya was a gesture of Imperial family recognition of which the JBS continues to be proud to this day.[26]

In the interim years, the JBS has successfully raised funds for or otherwise coordinated the construction of monuments at various sites throughout Honshu and southern Kyushu associated with major events in the Jinrai unit's war record. JBS members, both individually and collectively, continue to be among the Japanese war veteran interview subjects most eagerly sought after by researchers and journalists, and the group was even the subject of a British Broadcasting Corporation documentary in 1995. While the members are hesitant to disparage publicly the cause for – and circumstances under – which their comrades died in 1944–5, and they are impassioned in their desire that younger generations of Japanese know about and, if possible, honour these sacrifices, their testimony is notably free of patriotic bombast, and if now somewhat more restrained than it was 54 years ago, it is still in its own way as frank and honest as that displayed in the remarkable 1952 *Hagoromo Kai* publication. Honesty, realism and accountability in future *kamikaze* remembrance discourse will suffer for their absence when the last of the JBS members are gone in ten or twenty years.

The TZ: A case of 'top-down' memorialization

Since the dark days when *tokkō kuzure* was still a Japanese household word and intellectuals such as Shiga Naoya were pontificating about the social dangers of irreconcilable *kamikaze* veterans, the TZ has been working hard to nudge the *kamikaze* discourse in a distinctly hagiographic and culturally reaffirming (which is to say, revitalizationist) direction. Although, as noted in earlier sections, new membership among younger Japanese is increasing due to effective canvassing of college campuses and bookstores, memorial sites such as Yasukuni Shrine and certain history museums, and also through Internet activity, the TZ's core membership remains people in their late seventies, eighties and nineties who had direct personal experience with the *kamikaze* programme during the war, either as bereaved family members of *kamikaze* pilots killed in action, or as former *kamikaze* personnel who survived the war because of mechanical failure, sheer luck or simply because the war ended before they were given attack orders. The organization's leadership at present is roughly balanced between these two influentially dominant groups, although one up-and-coming probable successor group is comprised of recently retired Japan Self-Defence Force (JSDF)officers, who already show signs of lining up to receive the baton. To wit, at present, several recently retired JSDF officers are currently employed as full-time administrative staff for the organization, working at its head office in Minato-ku, Tokyo.

The TZ has its origins in the immediate post-war activities of five former Japanese admirals and generals who sent (or at least rubber-stamped the sending of) thousands of young men to their deaths either as Imperial General Headquarters staffers or at the head of army – and fleet-sized *kamikaze* commands during the war, but who, unlike many of their flag-ranked peers, chose not to make the ultimate gesture of contrition upon their nation's surrender.[27] Rather than committing suicide, as harsher interpretations of modern day *bushidō* might have proscribed for such a situation, these former officers instead chose to bear their collective cross of survivor's guilt by forming an association to memorialize *kamikaze* dead in 1952, shortly after the end of the GHQ Occupation of Japan made such activity legally possible.

The first official act of the group was to purchase twin bronze statuettes of the Buddhist deity Kannon-sama – one to commemorate Navy and the other to commemorate Army *kamikaze* dead – from a commercial war memorial enterprise selling religious statuary that had been organized by several influential Buddhist monks several months earlier.[28] In 1955, after several years of fruitless searching for a home temple for the statues, the memorial group – still calling itself at this point the 'Association to Venerate the *Kamikaze* Peace Kannons' (*Tokkō Heiwa Kannon Hōsan Kai*, hereafter *Hōsan Kai*) – was able to secure space on which to erect a special chapel for *kamikaze* memorialization at Setagaya Kannon temple in Tokyo. This location remains what could be considered the 'spiritual home' of the TZ to this day, and is also the venue of most of its most important ceremonies.[29]

The TZ attained an entirely new level of influence and organizational vigour under the stewardship of Takeda Tsuneyoshi, a first cousin of Emperor Hirohito and former imperial prince (Prince Takedanomiya) until the *miyake* 'cadet' branches of the Imperial family lost their royal status by GHQ fiat in 1947. Although Takeda had no personal connection to wartime *kamikaze* operations, he replaced former General Sugawara Michiō to become TZ chairman in 1981, a position he held until his death 11 years later.[30]

In addition to the significant prestige Takeda afforded the TZ as a former member of the imperial nobility, he also brought sturdy practical leadership skills to the job. Takeda was well qualified for the leadership of large organizations, having honed his skills during a wartime army career spent mostly with Imperial General Headquarters in Tokyo, and during his subsequent post-war incarnation as non-profit organization (NPO) board member extraordinaire, most notably as the Chairman of the Japan Olympic Committee (1962–9) – during which tenure he oversaw the Summer Olympics held in Tokyo in 1964 – and as a member of the International Olympic Committee (1967–81).[31] During Takeda's tutelage as TZ chairman, the activities of the group expanded nationwide and membership numbers soared, in large part due to the superlative personal contact network the ex-prince was able to use to foster extensive organizational cross-pollination with other veterans' and war memorial groups. Although Takeda passed away in 1992, the Imperial family's intimate connection with *kamikaze* memorialization continues to the present, as evidenced by the loyal attendance of Hirohito's youngest brother Prince Mikasanomiya at the Joint Memorial Ceremony for *Kamikaze* Personnel (*Gōdō Ireisai*), held in conjunction with other memorial organizations in the inner sanctum ceremonial space of Yasukuni Shrine during cherry blossom season each spring.[32]

Upon Takeda's death, the next chairman was Sejima Ryūzō, an even more fascinating historical character who can lay a solid claim to being the most influential pre-war Japanese policymaker currently living. As a brilliant young army officer during the war, Sejima was a staff officer with the Kwantung Army in Manchuria and China. While serving at Imperial General Headquarters (in the same section with Takeda Tsuneyoshi),[33] he was responsible for planning and overseeing the successful evacuation of Guadalcanal in 1943, and his day-to-day activities and decisions affected tens or even hundreds of thousands of lives. In 1945, back with the Kwantung Army staff, he was captured with several hundred thousand other Japanese soldiers when the Red Army rolled through Manchuria in the final days of the war. After enduring 11 years of internment in a series of Soviet gulags in Siberia and on the outskirts of Moscow, Sejima was finally released in 1956. Upon his return to Japan, he assumed his post-war guise as executive and eventual president and chairman of the board of the trading conglomerate now known as the Itochū Corporation.

In this and other key roles, Sejima has been one of the most influential crafters of modern Japanese business strategy, as well as an *eminence grise* in conservative politics.[34] In addition to countless other public and commercial leadership capacities, he has served as the chairman of the Japan Forum for Strategic Studies, and as an adviser to the president of NTT (the Nippon Telegraph and Telephone company). On the international stage, Sejima also exercises the application of Japanese soft power in his role as chairman of the Japan Art Association and the Kyocera Foundation, which sponsors the prestigious Kyoto Prize.[35]

The present head of the TZ, Yamamoto Takuma, former chairman of the board and current chairman emeritus of Japanese electronics and IT giant Fujitsu, assumed his post when Sejima stepped down as chairman for health reasons in 2004 at the age of 93. Yamamoto is the first *izoku* or 'war-bereaved' chairman of the TZ, his older brother having perished in 1944 in one of the first Army *kamikaze* missions. Chairman Yamamoto is overseeing the association at a critical juncture in its history. In his state-of-the-organization address on 30 March 2005 on the occasion of the annual *Gōdō Ireisai* Yasukuni rites,[36] Yamamoto stated that there are currently about 3600 dues-paying, full-fledged members of the TZ. Although 230 members died during the previous year, the addition of 1347 new members during the same period has pushed the finances of the association into the black for the first time in several years, putting the membership rolls comfortably above the 3000-member mark it needs to maintain its full-time staff and office.[37]

While the majority of the TZ's roster still consists of war-generation members, their numbers are dropping quickly, and it is unrealistic to expect that their dominance will last far into the 2010s. This demographic reality forebodes changes ahead for the group, particularly in terms of the theological orientation of its memorial mission. But unlike the JBS, it is evident that the TZ has no intention of fading away like Douglas MacArthur's proverbial old soldier when the last *kamikaze* veterans are gone: when the TZ membership is eventually and inevitably limited to individuals unable to mourn the *kamikaze* dead as lost loved ones, then the group will likely be forced to cling to life by accelerating its ongoing transformation from memorial association to political organization. In this sense, it is possible to view the group's recent recruiting efforts, focusing on young people with no personal connection either to the *kamikaze* or to the war, as indicating that contingency planning along these lines is already underway.

Whither *kamikaze* memorialization?

The fact that popular culture treatments of *kamikaze*-themed material have enjoyed perennial currency with Japanese audiences virtually since the end of the war (with an aspect of cyclical waxing and waning) can in itself be considered a qualified success for revisionists with a *kamikaze* hagiography agenda, as well as an indication of the special and considerably rehabilitated place *kamikaze* veterans hold in the hearts and minds of a large section of the Japanese public. Nevertheless all of this falls far short of the *kamikaze* pilot statues overlooking public parks and playgrounds and chapters on *kamikaze* in school textbooks that revisionists dream of making a reality someday.

It must be acknowledged that the entertainment value of the occasional titillating stimulation of ethnocentric pride is one matter, that permanent and official public memorialization is quite another, and that the two genres should not be confused. Entertainment is a matter of private consumption, but school curricula and monuments in public spaces must be lived with, regardless of personal preference. At present, there are still too many Japanese who would be too discomfited by intrusions of officially sanctioned *kamikaze* lore into their (or perhaps more critically, their children's) lives for memorial aspirations along the lines envisioned by revisionists to become a reality.

Many in the revisionist community blame GHQ policies of the 1940s for this situation. Others blame the self-excoriating modern history education curricula of the 1950s and 1960s that were in part legacies of those Allied occupation-enforced policies. Still others see the culprit as the political apathy that has accompanied Japanese society's headlong rush into material consumerism from the 1970s on. But the assignment of blame cannot alter the likelihood that in terms of revisionism's desire to establish a heroic 'Lost Cause' hagiography for the Asia-Pacific War that would be an integral part of Japanese culture and society[38] – an aspiration both symbolized and spearheaded by ongoing attempts to rehabilitate the historical legacy of the *kamikaze* – the window of opportunity may have long since closed, if, in fact, it was ever open at all. The revisionist engine that worked up such a high head of steam in the 1990s, with vigorous encouragement by many Japanese veterans, may turn out, in the final verdict, to have been a case of too little, too late. The veterans – in good measure out of contrition for their defeat – were simply silent for too long.

Reversing this state of affairs will be a daunting task for revisionists, and nothing less than impossible without the cooperation of major players in media, education and public policy. But that said, given the already multi-generational nature of post-war Japanese revisionism, the eventual fruition of its aims at the hands of a younger breed of acolytes after patiently executed institutional marches through corporate media hierarchies, the Japanese academy and bureaucracies such as the Japan Self Defence Forces or the Ministry of Education and Science is not unthinkable.

Moreover, in the meantime and for the foreseeable future, China will continue to grow stronger, not only militarily and economically but also in terms of regional and even global cultural influence. It is also likely that de facto Chinese vassal state North Korea will continue and, perhaps, escalate its attention-getting provocations of Japan. These developments may yet provoke enough anxiety and resentment over the plight of a US-protected Japan in slow decline relative to Chinese regional hegemony to push the Japanese to a cultural crisis consciousness tipping point of some sort. If this occurs, it is conceivable that everything the *kamikaze* hagiographers have worked 60 years to accomplish could fall into place virtually overnight. If the Japanese have demonstrated any unique collective characteristic over the centuries, it is the ability, ideologically speaking, to stop on a dime and change societal course with breathtaking speed and unity of purpose in response to perceived external threat. Considering this quality, a society-wide Japanese revitalization movement invoking beatifically beaming ghosts in army khaki, navy blue and *kamikaze* pilot brown someday in the near future is a contingency that may be less than likely, but it is still far from impossible.

Notes

Names in the body of this chapter are according to the Japanese convention of giving the surname and then the given name. Japanese names elsewhere in the book are presented according to the Western convention.

1. These are figures for casualties officially recognized as tokkō deaths by Yasukuni Shrine and by the *Tokkō Zaidan*. Counts including crew losses from the final and fatal 'maritime tokkō' (*kaijō tokkō*) sortie of the battleship *Yamato* from Kure in April 1945 put these numbers into the 8000s.
2. M. G. Sheftall, *Blossoms in the Wind: Human Legacies of the Kamikaze* (New York: New American Library, 2005), p. 433.
3. 'Revitalizationalist movement', a term first used by anthropologist Anthony F. C. Wallace in 1956, is a key concept in the methodology employed by Gaines M. Foster in his *Ghosts of the Confederacy: Defeat, the Lost Cause, and the Emergence of the New South* (New York: Oxford University Press, 1987), applied in the context of post-bellum ex-Confederates who participated in what Foster likens to a kind of 'Ghost Dance' of defeat denial (cf. pp. 47–62). Foster refers to participants in this movement as 'revitalizationists', and I use Foster's

term to categorize a similar identity and orientation for (some of) the discourse participants examined in this study.

4. See Sven Saaler, *Politics, Memory and Public Opinion: The History Textbook Controversy and Japanese Society* (Munich: IUDICIUM, 2005).
5. See Sheftall, *Blossoms in the Wind*, p. 358. It should be noted that Yasukuni extends this apotheosis to all Japanese (or as the case may be, colonial 'proxy Japanese') troops who died in the war, although the preponderance of *kamikaze*-themed statuary and other artwork at the shrine (much of it made possible through TZ and JBS fundraising efforts or direct donation) makes it clear that the *kamikaze* pilots are considered the 'elite of the elite' in Yasukuni's pantheon of nation-protecting deities (*gokoku no mitama*).
6. TZ chairman Yamamoto Takuma pinpoints the origins of these undesirable trends not in the immediate aftermath of defeat, but rather, ten years later, at the onset of the Period of High Economic Growth (*kōdo seichōki*) (Yamamoto Takuma, keynote address at Joint Special Attack Memorial Service, 30 March 2006, Yasukuni Shrine, Tokyo).
7. This is the basic gist of the JBS's organizational goal as stated in the introduction to Ōka (Hagoromo Kai, 1952), and as echoed by the JBS's lifetime unofficial chairman, Suzuki Hideo in an interview on 21 March 2006.
8. Accordingly, members of both groups have little tolerance for comparison of twenty-first century suicide bombing tactics with Japan's wartime *kamikaze* programme.
9. See Obinata Sumio,Yamashina Saburō, Yamada Akira, and Ishiyama Hisao, *Kimitachi wa sensō de shineru ka: Kobayashi Yoshinori 'sensōron' hihan* (Tokyo: Ōtsuki Shoten, 1998) and *Datsu sensōron: Kobayashi Yoshinori to no saiban wo hete* (Uesugi Satoshi, ed.) (Osaka: Tōhō Shuppan, 2000). Shida Masamichi, *Saigo no tokkōtai'in: nidome no 'isho'* (Tokyo: Kōbunkyū, 1998).
10. See Ivan Morris, *The Nobility of Failure: Tragic Heroes in the History of Japan* (New York: Holt, Rinehart & Winston, 1975).
11. A very accessible English translation of this seminal book has been written by Midori Yamanouchi and Joseph L. Quinn, SJ as *Listen to the Voices of the Sea* (Scranton: University of Scranton Press, 2002).
12. Popular Japanese works that can be considered representative of a genre of 'kamikaze encomium' are comic book artist Kobayashi Yoshinori's three-volume *Sensōron* series (Tokyo: Gentōsha, 1998–2003), Kudō Yukie's *Tokkō no Rekuiemu* (Tokyo: Chūōkōronsha, 2001), and the films *Winds of God* (Shōchiku, 1995), *Kimi wo Wasurenai* (Nihon Herald, 1995), *Hotaru* (Tōei, 2001) and *Otokotachi no Yamato* (Tōhō, 2005).
13. Anthony F. C. Wallace, 'Revitalization Movements', *American Anthropologist 58* (1956), pp. 264–81, cited in Foster, *Ghosts of the Confederacy*, p. 56.
14. Ralph Linton, 'Nativistic Movements', *American Anthropologist 45* (April–June 1943), pp. 230–40, quotation from p. 233. Cited in Foster, *Ghosts of the Confederacy.*
15. John Nathan, *Japan Unbound* (New York: Houghton Mifflin Company, 2004), cf. pp. 119–67; Yoshida Yutaka, *Nihonjin no Sensōkan* (Tokyo: Iwanami Gendai Bunko, 2005).
16. See *Asahi Shimbun*, 6 August and 15 August 1945 (particularly the editorial pages of the latter edition). These and all other English translations of Japanese text are the work of the writer, unless otherwise noted.

17. General Headquarters of the Supreme Commander of the Allied Powers, that is, the Allied military occupation administration of General Douglas MacArthur.
18. 'Voice' (*Koe*) Column, *Asahi Shimbun*, 16 December 1945.
19. 'Voice' (*Koe*) Column, *Asahi Shimbun*, 13 January 1946.
20. Ibid.
21. Ango Sakaguchi, 'Discourse on Decadence', translated by Seiji M. Lippit, *Review of Japanese Culture and Society*, I (October 1986), 1–5; cited in John Dower's *Embracing Defeat: Japan in the Wake of World War II* (New York: W.W. Norton, 1999), p. 156.
22. Dower's *Embracing Defeat* provides an excellent explanation of this psychological state. Alexander and Margarete Mitscherlich, *The Inability to Mourn* (New York: Grove Press, 1975), also note an intriguingly similar phenomenon occurring in post-Second World War Germany.
23. The existence of the GHQ guideline to which Suzuki refers has been confirmed by numerous other *kamikaze* veterans during my five years of field work researching *kamikaze* groups, but I have as of yet been unable to find the original document.
24. Both Army and Navy service academy graduates, viewed by the GHQ as recalcitrant militarist risks, were under proscription in most aspects of public life at the time, and most kept a low profile for the duration of the Occupation.
25. Naitō Hatsuho, *Gokugen no tokkōki: Ōka* (Tokyo: Chūō Kōron Shinsha, 1999), p. 302. The blooming of the first blossom on this tree each year is a major news event in Japan.
26. Yunokawa Masamori and Shinjō Hiroshi (eds), *Kaigun Jinrai Butai* (official unit history, non-commercial publication) (Tokyo: Kōgakusha, 1996), p. 71.
27. These men were former Prime Minister (August–October 1945) and *miyake* prince, General Higashikuni Naruhiko (Imperial Military Academy Class of 1908); former Navy Minister and Chief of Staff, Admiral Oikawa Koshirō (Imperial Naval Academy 1903); former Chief of Army Air Operations, General Kawabe Shōzō (IMA 1906); former commander of the Sixth Air Army, Lieutenant General Sugawara Michiō (IMA 1908); and former commander of the Third Air Fleet, Vice Admiral Teraoka Kinpei (INA 1912). Service academy class year and last major wartime command information for this group is from Hata Ikuhiko, *Nihon Rikukaigun Sōgō Jiten* (Tokyo: Tokyo Daigaku Shuppan Kai, 1991).
28. Interview with Setagaya Kannon abbot Ōta Kenshō, 11 February 2006; see also *Kaihō Tokkō*, vols 4 and 59.
29. Ibid. An additional *Hōsan Kai* contribution to *kamikaze* memorialization that continues to influence the discourse to this day, and to a degree far beyond the TZ's present reach, was the key generative role the group played in organizing post-war *kamikaze* memorialization activities at Chiran, Kagoshima Prefecture. The resulting tourist attraction/memorial facility now receives nearly a million visitors a year.
30. See http://www.tokkotai.or.jp/enkaku.html. This page is part of the official TZ website.
31. *Olympic Review*, vol. 296, June 1992, p. 291.
32. I have personally witnessed the prince's attendance at these ceremonies from 2002 to 2005. Mikasanomiya, 91 years old at the time of writing of this chapter, did not attend the 2006 event.

33. *Kaihō Tokkō*, vol. 64.
34. http://www.asahi-net.or.jp/~ny3k-kbys/contents/digi7.html.
35. Sejima Ryūzō, *Ikusanga: Sejima Ryūzō kaisō roku* (Tokyo: Sankei Shinbunsha, 1996), pp. 667–8. Recent Kyoto Prize laureates include immunologist Leonard Herzenberg, designer Issey Miyake and philosopher Jürgen Habermas.
36. The *Gōdō Ireisai*, or 'Joint Memorial Ceremony', conducted at Yasukuni Shrine every year, is sponsored under the auspices of several veterans' and memorial groups, including the *Tokkō Zaidan*, which is responsible for organizing the event.
37. Yamamoto Takuma, speech at reception after *Gōdō Ireisai*, 30 March 2005.
38. This is as opposed to the more pitiable-than-heroic cult of victimization surrounding Hiroshima, Nagasaki and the fire raids, which of course has already achieved such permanent iconic status. See James J. Orr, *The Victim as Hero: Ideologies of Peace and National Identity in Postwar Japan* (Honolulu: University of Hawai'i Press, 2001) and John Whittier Treat, *Writing Ground Zero: Japanese Literature and the Atomic Bomb* (Chicago: The University of Chicago Press, 1995).

11
Confederate Defeat and Cultural Expressions of Memory, 1877–1940

Karen L. Cox

One cannot speak or write about the post-bellum American South without understanding that Confederate defeat is at the core of its collective memory. For southern blacks, defeat meant emancipation from slavery, yet for southern whites defeat brought devastation on a number of levels – economic, social, and psychological. The failed attempt by southern states to form an independent nation, the trauma of defeat, the occupation of Federal forces during Reconstruction, and the economic and racial upheaval brought on by the end of slavery sent white southerners into a tailspin. They came to terms with defeat by creating a mythological narrative, commonly referred to as the 'Lost Cause'. This version of events cast Confederate veterans as heroes, its leaders as martyrs, and the Civil War as a battle for states' rights, not the preservation of slavery.

Historians who have studied Confederate defeat's impact on southern identity and memory fall into two categories: those who analyse Lost Cause ideology and its impact on the southern psyche, and those who have examined the individuals and organizations that were instrumental in perpetuating the mythology. Charles Reagan Wilson's *Baptized in Blood* identified the Lost Cause as a 'civil religion' wherein white southerners came to terms with defeat through religious justification of the southern cause; in other words, the cause was just, and in God's eyes, the South was right. Gaines Foster's work *Ghosts of the Confederacy* defined Lost Cause activities of Confederate veterans as a 'celebration' of the 'Confederate tradition'. Again, white southerners came to terms with defeat by developing an ideology that allowed them to accept that while odds were stacked against them, southern soldiers fought valiantly and their cause was justified by their defence of the Tenth Amendment to the United States Constitution protecting the rights of states – even if that right was to maintain the institution of slavery. My own work on the United Daughters of the Confederacy (UDC) explores women's role in preserving the ideology of the Lost Cause for future generations. Yet it is clear that the Lost Cause permeated southern culture beyond what has heretofore been written.[1]

One of the great ironies of Confederate defeat is that failure to create a separate southern nation turned into a celebration of the defeated and the virtues of southern nationalism. An entire culture, in fact, emerged to embrace the Confederacy and its heroes – a culture that shaped the New South from the late nineteenth century until World War II when the region's economic, social, and political landscape radically changed. Confederate culture, which included a longing for the Old South, helped to shape white southern identity around the region's past – a useable past – that emphasized dedication to the Confederate cause and the culture of the antebellum South. Confederate organizations of men, women, and children were formed to perpetuate the Lost Cause, but the extent of this Confederate culture did not stop with its defenders. Southern educators embraced this culture by placing Confederate flags in the classrooms of public schools and establishing curricula that incorporated the study of Confederate heroes. Southern whites also founded museums and archives to exhibit and preserve Confederate relics and manuscripts.

The road to reconciliation

To understand how the cultural memory of defeat developed in the South, one must understand the evolution of the region after the Civil War as well as its changing relationship with the North. The means by which white southerners emerged from defeat was to embrace the ideology of the Lost Cause, which allowed them to honour the defeated in the immediate aftermath of the war. After Reconstruction and the return of home rule, the Lost Cause was used to celebrate Confederate veterans as heroes and the cause of southern nationalism. The Lost Cause phenomenon of the late nineteenth century helped to perpetuate the mythology of the Old South and proved to be a strong bulwark against modernity and a New South based on industrialization and manufacturing.

The South's relationship with the North also went through significant changes in the decades following the Civil War. White southerners resented Reconstruction and deplored any rights of citizenship granted to blacks; furthermore, they considered military occupation an insult added to injury. Although the 'Tragic Era' remained a significant part of Confederate memory, when it ended in 1877 and former Confederates restaked their claim to govern the South, the road to reconciliation with the North appeared within reach and white southerners were receptive to forging new bonds with their northern brethren.

There were also many northerners who, at war's end, wanted a swift reconciliation with the South, and that sentiment strengthened in subsequent years. As David Blight argues, the North's interest in reconciliation with the South overshadowed one of the primary outcomes of the war – emancipation – as shoring up a brotherhood based on Anglo-Saxon supremacy overshadowed the enforcement of Reconstruction amendments. More importantly, the South represented an enormous market for northern businesses. The North's historical amnesia, moreover, also meant that when it came to the way southern governments treated their black citizens they were free to do as they pleased. The result, according to Blight, was that within the politics of Civil War memory, the Northern reconciliationist view blended with the South's white supremacist view of the Civil War to eclipse the black memory of the war as about emancipation. By accepting the South's historical revisionism, northerners met a major condition for reconciliation with the white South. This paved the way for what Nina Silber calls the 'romance of reunion' expressed through gatherings of the Blue and the Grey as well as in popular culture.[2]

Cultural reconciliation

In addition to the social reconciliation that took place between northern and southern whites, there was a cultural reconciliation between the regions that can be understood through the production and consumption of consumer goods in the rapidly changing economy of the country during the last quarter of the nineteenth century – change that saw the emergence of southern memory in the marketplace. Most significantly, the change from an economy based primarily on agriculture to one based on production and consumption led to changes in the marketplace. Indeed, the end of the nineteenth century marked the era of the national mass market. Urbanization contributed to this change, as did the significant growth of the national rail network. According to historian Richard Tedlow, 'the completion of the new transportation and communication infrastructure made it possible to distribute goods nationally.'[3]

New technology created machines that were faster and could produce thousands of units of a product, from cigarettes to breakfast cereals. Production was possible on a scale never before imagined, as were the accompanying profits. This type of production helped meet consumer demands in far-flung markets, and created consumers for a variety of luxury goods. Mass production contributed to another important nineteenth-century development – mass marketing and the development of brands.

The South was not immune to these changes and consumer culture in the New South also reflected the popularity of the Lost Cause. Businessmen in the region (and outside of it) recognized the profitability of marketing consumer goods as products of 'Dixie'. A market also existed for the publication of pro-southern textbooks that justified the Confederate cause and romanticized the Old South. Consumers of Confederate culture included both individuals and the state. Indeed, southern state and local governments were significant consumers of the Lost Cause as they supported the creation of Confederate holidays, helped fund Confederate monuments, made appropriations to Confederate museums, and made pro-southern textbooks required reading for students in the region's public schools. The result was a culture that emphasized a mythic past, a society that invested literally as well as figuratively in the Lost Cause, and a region critical of modernity. Even proponents of the New South who welcomed industrialization, such as *Atlanta Constitution* editor Henry Grady, understood that they were engaged in an uphill battle against a society that revered its agrarian traditions.[4]

In fact, despite the enthusiasm of New South spokesmen about the region's natural resources and the potential for northern investors, what becomes clear is that white southerners on the whole were less interested in following the path of Northern industrialization and modernization in favour of recreating the New South on the basis of the Old. The most influential narrative for citizens of the region was not the 'New South Creed' of which Paul Gaston has so eloquently written, but the Lost Cause mythology that still regarded southern nationalism as a worthwhile endeavour. The means by which white southerners sought to pursue that goal was to pay homage to the Old South, the culture that was to be preserved by the Confederacy. Thus, the 'Lost Cause' was not entirely lost.[5]

One of the earliest means by which white southerners sought to preserve Confederate culture was through literature and in this, they found support among northern readers. Throughout the nineteenth century, plantation literature was one of the most popular genres of literature in the country. Following the Civil War that literature found new audiences in the North and Midwest due in large part to the literary industry whose base was in the North. J. B. Lippincott of Philadelphia, Pennsylvania, and E. P. Dutton & Company in New York, for example, maintained a large list of southern history and literature titles that romanticized the region.

Readers throughout the nation found in this literature a South that was not simply the place of rebellion, but a region where the pace and way of life provided welcome contrast to the urban and industrialized regions in which they lived. Thomas Nelson Page's plantation novels and Joel Chandler Harris's Uncle Remus stories perpetuated the myth of the 'sunny' South, helping to deflect criticism from contemporary race relations in the late nineteenth century. Indeed, these and many other works of fiction played a critical role in persuading the North that the South knew best how to address its 'Negro problem'.[6]

Revisionist histories of the Civil War were as significant to southern readers as plantation literature was popular among northern readers. Members of the United Confederate Veterans (UCV) and the UDC had long railed against what they perceived to be 'falsehoods' perpetuated by Northern writers of the war, and members engaged in the promotion of literature that told the 'truth' of the Confederacy and the Old South as they understood it. To combat 'biased' histories, both organizations encouraged and supported the creation of pro-Confederate histories. The result was that a number of southern publishers capitalized on this gap in the marketplace and responded by issuing lists of pro-Southern literature.[7]

One of these companies was the B. F. Johnson Publishing Company in Richmond, Virginia. Benjamin Franklin Johnson entered the publishing business in 1876 at the age of 21, as a partner in the firm of Mosby and Johnson. By 1884, Johnson became the sole proprietor of the publishing house. Initially booksellers of standard religious works and Bibles, by 1900 Johnson boasted of his efforts to 'build up a great Educational Publishing House in the South'. In fact, Johnson was quite successful and many of the books published by the firm were the official selections by the State of Virginia for its public schools.[8]

B. F. Johnson's entrepreneurial spirit was, on some level, at odds with what he proposed to sell – a book that touted the Confederate past. That is, Johnson used the tactics of modernity – marketing and advertising – to promote a book written in the anti-modern tradition of the Lost Cause. In an advertisement for *The Story of the Confederate States*, published in 1895, Johnson plays on the sympathies of southern readers to sell the book. 'Unfortunately most of the history has been written by the other side, and the Southern people have been placed at a disadvantage,' Johnson wrote. 'Those who are fortunate enough to secure it [the book] will rise from the reading inspired with proper pride in their SOUTHERN LAND, and with reverence for their gallant ancestors.' In case people should think otherwise, Johnson warned that this was no 'make-shift' book 'gotten up to wring a few dollars out of a credulous public'. In fact, Johnson was trying to 'wring a few dollars' out of the public, and was very successful at doing so through the publication of several texts that were written by southerners for southern audiences and that often portrayed Confederate soldiers as heroes and the Confederate cause as just.[9]

Johnson's story illustrates the earning potential associated with selling Confederate memory, and books were just one of several means to profit from the sale of the Lost Cause. Moreover, southern businesses were not alone in their efforts to tap into this market. Northern entrepreneurs – book publishers, producers of consumer goods, and music publishers – also recognized the profitability of employing the Dixie brand. North and South, therefore, businessmen sought to capitalize on the marketing and consumption of Dixie that incorporated the mythology and traditions of the southern past.

Literature, therefore, as well as other forms of consumer culture associated with the Lost Cause fit into larger trends of consumption during the late nineteenth and early twentieth centuries. The rise of mass consumerism, as described by historian T. Jackson Lears, was a reaction based on the general anxiety that accompanied modernity, and that people bought into a culture with which they felt comfortable and familiar. The anxiety about the modern world was expressed emotionally and religiously, contributing to what Lears calls a 'therapeutic ethos' in all aspects of life, including consumerism. Advertisers capitalized on this anxiety by creating advertisements that offered what Lears calls the 'psychological impetus for the rise of consumer culture'.[10]

According to political theorist Michael Walzer, consumers respond to anxiety by purchasing what he calls 'social goods'. These objects, regardless of time and place, help us identify the 'social world to which we feel we belong'. This is not only a practical reality but also a political one, because such goods also serve as markers of citizenship, in other words, a sense of belonging. Belonging in a consumer society, and a society reacting to change, is not simply about the commodity, but about the access to the world that that commodity represents. So, how did the 'social goods' of Confederate memory make its way into the marketplace of the late nineteenth and early twentieth centuries? The answer lies in the emergence of modern advertising.[11]

The emergence of modern advertising

For most of the nineteenth century, the majority of advertising appeared in newspapers, and when advertisements appeared in literary magazines they were usually confined to the rear of the journal because of literary editors' general disregard for advertising. Religious magazines, the most popular magazines of the mid-nineteenth century were the exception, and profited handsomely from advertising stimulated by the contemporary controversies between fundamentalism and the Social Gospel Movement, as well as the publication of Charles Darwin's *On the Origin of the Species*.[12]

Patent medicines and insurance companies were often the leading advertisers in nineteenth-century newspapers and magazines, and continued to be important sources of advertising revenue throughout the nineteenth century. Yet, between 1865 and 1900, advertising assumed a more important role in the marketplace, which corresponded to the rise of mass production and consumption, as well as the expansion of popular magazines. By the end of the century, popular magazines replaced religious and literary magazines in circulation, spurred on by modern advertising. That advertising, aimed at selling luxury goods to middle-class consumers, was the direct result of the rise of mass production.[13]

By 1900 businesses developed marketing strategies to facilitate mass consumption, which meant changes in product advertising. As one scholar puts it, 'advertising came to be the major communications link between mass production and mass consumption.' Advertising, which for most of the nineteenth century did little more than explain what the product was, now incorporated slogans and messages to attract consumers. More importantly, manufacturers sought to sustain their customer base by creating a 'super name – a brand'.[14]

Before there were brands, soap was soap; with the creation of brands, 'Ivory' or 'Pear's' immediately conveyed soap and the qualities assigned to it by their manufacturers. The rise of brand-name advertising proved effective, as consumers increasingly depended on brands to assure them of 'real or perceived quality', and brand-name advertising rapidly expanded in the early twentieth century.[15]

Branding Dixie

Branding and consuming Confederate memory can best be understood through an examination of advertising as a medium that can both reflect and reinforce cultural values. Richard Pollay, who has written extensively on the history of advertising, argues that 'whether advertising imagery reflects the present or the future, it supports an image of society, and this is enough to warrant our attention'. Certainly not all cultural values of a society are reinforced; however, there remain social and cultural implications related to those values that *are* conveyed through advertising. This is particularly evident in the way advertisements branded Dixie.[16]

What did it mean to brand Dixie? How was the South branded and what was conveyed in advertising in the period between 1890 and 1930? As used here, 'Dixie' serves as both symbol and brand. The actual word 'Dixie', when used in the message of an advertisement symbolized something much broader than a geographic region. It conveyed ideas about landscape and Confederate memory, race and class, the rural versus the urban, agriculture versus industry, and anti-modern versus modern – and seen often through the lens of the Confederate past. The Dixie brand was, as historian Grace Hales has written, 'a regional identity made or marketed as southern'. When used as part of a product name, as in 'Dixie Cigarettes' or 'Dixie Nerve and Bone Liniment', it often (though not always) meant the product was manufactured in the South.[17]

The cultural meaning of Dixie, moreover, varied between regional and national publications. While regional magazines were an obvious place to brand Dixie and employ the rhetoric of the Lost Cause and the Old South, national advertisers also employed those messages. They capitalized on northern sentiments about reconciliation with the South. Thus, branding Dixie might mean images of a plantation or a black servant on a product to lull northerners into the idea of an idyllic region. Such advertising also came at a time when southern values on race had traction, as northern whites reacted negatively to the influx of immigrants from Southern and Eastern Europe.

Branding Dixie in the South was a relatively simple proposition. The market for all things Dixie was healthy at the turn of the twentieth century. Sales of monuments and pro-Confederate books, for example, allowed regional businesses to create new consumers of Confederate products by branding those items as having been made in the South or using iconography for which there was instant recognition among southern consumers. Martial icons such as the Confederate battle flag or an image of Robert E. Lee, for example, were often used in advertisements and instantly conveyed messages to consumers about regional identity and values. The Ku Klux Klan was also marketed to southern consumers via popular literature and, eventually, the film *Birth of a Nation*. Yet D. W. Griffith's epic film based on Thomas Dixon's novel,

The Clansmen, appealed to northern audiences as well. Again, the shared cultural values of white supremacy and racism were sold to both northern and southern audiences.

Icons of the Old South, especially the 'mammy' who often appeared as a large black woman wearing a bandana, appeared with less frequency in regional advertising but were favoured by northern manufacturers who advertised nationally.[18] Southern businesses were more likely to employ images of industry and Confederate heritage during this period, as southern race relations were not conducive to romanticized images of blacks in advertising, even though the 'uncle' and 'mammy' were thought by southern whites to serve as ideal role models for the younger generation of blacks. Northern publishers and businesses, on the other hand, marketed to a broader national audience that bought into these stock images of docile and happy black servants, icons of the Old South, because they represented a life of leisure to consumers whose lives in the urban-industrial North were anything but leisurely.

For the purposes of this chapter, the discussion of branding Dixie within the region is limited to the advertising found in the *Confederate Veteran*, a regional publication with a circulation of around 20,000 copies per month and was published between 1893 and 1932. It is an obvious choice for examining the advertisement of Confederate culture, and it is also useful for identifying regional businesses as well as national advertisers who recognized a niche market for Confederate memorabilia among consumers of the magazine.[19]

Advertising in the *Veteran* (the shorthand used by contemporaries) reflected trends in modern advertising. Just like national magazines, the *Veteran* hosted advertisements for patent medicines and insurance companies, railroads and private schools, and domestic goods that appealed to women – increasingly recognized by manufacturers as their most important consumers. What differentiated advertisements placed by southern businesses, as compared to those of the North, was the link to Confederate memory, either through language or regional icons.

Among the regular advertisements of patent medicines was Dr G. H. Tichenor's 'Antiseptic Refrigerant for Wounds'. Tichenor's medicine was manufactured in New Orleans and its label incorporated a small group of Confederate soldiers hoisting the battle flag. While the product itself claimed to heal wounds 'quicker and with less pain on man or beast than any compound known', it was vetted by the fact that George Tichenor was a Confederate veteran, the brand was pure Dixie, it was made in the South, and the perceived quality of the product was directly linked to an image of honour – heroic southern soldiers holding aloft the Confederate flag.[20]

Tobacco companies, and the South had several, were some of the earliest businesses to adopt advertising budgets. The American Tobacco Company advertised nationally, yet there were smaller regional companies who placed a Confederate spin on their advertisements. In 1906, Taylor Brothers of Winston-Salem, North Carolina, advertised its 'Stars & Bars' chewing tobacco by appealing directly to veterans and offered consumers, with the purchase of a plug of tobacco, a 'Stars and Bars calendar with all Confederate flags in colors, and a history of each flag'. Richmond, Virginia, tobacco manufacturers Allen & Ginter, although they never advertised in the *Veteran*, sold 'Dixie Cigarettes' and were leaders in using what was then a new advertising gimmick, collectible trading cards. By 1929, Edgeworth Tobacco, also of Richmond, used the new technology of radio to advertise. Also known as WRVA, the station's slogan was 'Down Where the South Begins'. In addition to promoting the link between its station and the 'capital of the Southland', the company promoted the links between its products and Confederate history in the Richmond area, telling its listeners 'our station endeavors to send to you the spirit of hospitality, graciousness and charm that characterizes the Southland'.[21]

Souvenirs, essentially luxury items, frequently appeared for sale in the *Veteran*. S. Thomas & Brothers of Charleston, South Carolina, sold 'Confederate buttons, souvenir spoons, pictures of Fort Sumter, President Jefferson Davis, Gen. Lee, Gen. Johnson, Gen. Gordon, [and] Gen. Hampton'. S. N. Meyer of Washington, DC, advertised 'Ladies' Belt Plates' with the raised letters 'C.S.A.' as a 'handsome and useful souvenir'. 'The Game of Confederate Heroes' was advertised as both a souvenir and a fundraiser. It was composed of 52 cards 'divided into thirteen books', no doubt a reference to the 13 states of the Confederacy. Confederate military and political leaders were illustrated in colour on the cards, and the proceeds of the game went to the Sam Davis Monument Fund in Nashville, Tennessee. As luxury items, these products used Confederate memory for appeal and the advertisements suggested that such items were, like Confederate 'relics', sacred in value.[22]

Increasing rail travel required that railroad companies advertise and, here again, Confederate memory was evident. The earliest railroad advertisements listed nothing more than times of service. This changed with the advent of additional rail lines and tourism travel. The 'Dixie Flyer' was regularly advertised in the *Veteran* as was the Cotton Belt

Route. The Illinois Central Railroad ran the Dixie Flyer along what was known as the 'Lookout Mountain Route', between Chicago, Illinois, and, Jacksonville, Florida. Travelling through Dixie on a train that carried the same name helped to brand the route. The Cotton Belt Route advertised to take veterans west to Texas for a reunion by offering its travellers a 'free picture of Gen. Lee' along with a copy of his farewell address 'suitable for framing'. In this instance, none other than Robert E. Lee – the South's most revered hero of the Civil War – is used to sell the rail service.[23]

Some products advertised in the magazine met specific market demands. Uniforms for attending reunions of Confederate veterans or of the Blue and the Gray represented a niche market for both northern and southern companies. M. C. Lilley & Company of Cincinnati, Ohio, and, after 1910, the Pettibone Company of Louisville, Kentucky, both advertised to fill the demand for reproduction of uniforms. As the ageing veteran population dwindled in number, the National Casket Company of Nashville, Tennessee, marketed to consumers with the message that 'Confederate Veterans should be buried in Confederate grey broadcloth-covered caskets'. Though not a necessity for burial, the company used the emotional appeal of the Lost Cause to sell its Confederate-themed caskets. The Dixie Artificial Limb Company fulfilled a real need among many Confederate veterans who lost their limbs during the Civil War. J. C. Griffin, who managed the Nashville-based company, advertised to its potential customers to get rid of their 'OLD PEG LEG', which caused them embarrassment in favour of his 'improved willow wood limbs'. For individuals considering a new limb, the idea that the manufacturer was based in Dixie may have afforded them some measure of comfort.[24]

Monuments and books were advertised in the *Veteran* with the most frequency between 1890 and 1930. Monuments were the most public symbols of Confederate defeat and memory between 1895 and World War I, as hundreds were placed on the southern landscape driven by the efforts of the UDC to honour their Confederate forbears. Thus, monument companies functioned in a ready-made market for their products. Two companies advertised regularly during those years – the Muldoon Monument Company of Louisville, Kentucky, and the McNeel Marble Company of Marietta, Georgia. Their advertisements made direct appeals to the UDC, the organization most responsible for erecting Confederate monuments. The Muldoon company claimed to be the 'oldest and most reliable' and responsible for 'nine-tenths of the Confederate monuments in the United

States'. The advertisements were often branded with an image of a Confederate soldier cast in stone, and the companies publicized having a history of working with the Daughters to build monuments throughout the South.[25]

Publishers of southern history and literature, more than any other company, filled the pages of the *Veteran* with advertisements for their books, just as they did in national literary magazines. Pro-southern and pro-Confederate literature was the order of the day in the early twentieth century, and both regional and national publishers sought to meet and create consumer demand. Publishers in New York, Philadelphia, and Chicago advertised in the *Veteran* along with B. F. Johnson of Richmond, Page Publishing of Baltimore, and the University Press of Nashville.

New York publishers were capable of publishing as much racist literature set in Dixie as were southern publishers, which begs the question: 'why were white southerners convinced that northern histories of the war misrepresented Confederate history?'. The answer may be twofold: literature condemning the Confederacy for rebellion did exist, but a second explanation may stem from the marketplace. From whom should southerners buy their books? B. F. Johnson or Scribner? Finally, there were publishing companies in New York, specifically the Neale Publishing Company, that were founded by white southerners with Confederate sympathies.

The New York publishing industry recognized the profitability of the Lost Cause by advertising its books in southern periodicals such as the *Veteran*. So, too, the Ohio uniform manufacturer that capitalized on the needs for the men attending veterans' reunions. Flag companies in New York and New Jersey filled consumer demand by advertising that they manufactured and sold Confederate flags and bunting. Branding Dixie, however, extended into the national marketplace as both northern and southern manufacturers associated the quality and value of their products with the qualities and values that northern advertisers, and subsequently consumers, associated with the South.

Several historians have written about the use of southern blacks in national advertising. In print advertising, and later, on radio and television, blacks were demeaned by the creation of stereotypical images. The black 'mammy' and 'uncle' caricatures appealed to white consumers nationwide, because the image often conveyed blacks as docile servants who willingly took care of their white employers. Furthermore, such 'faithful servants' projected the image of leisure to which white middle-class consumers aspired. Significantly, the advertisement campaigns that branded Aunt Jemima and Maxwell House (based on a real

Nashville Hotel) as icons of the Old South were developed by the prominent New York advertising agency, J. Walter Thompson.[26]

The Thompson agency successfully employed black stereotypes and 'southern style', and influenced other advertisers who used similar images in their own advertisement campaigns. New York publisher E. P. Dutton & Company advertised its book *ABC in Dixie: A Plantation Alphabet* (1904) by using a cartoon-like figure of an elderly black man in a top hat and tails – suggestive of the male house servants of the Old South. Indeed, the advertisement explains that the book offers 'verses and pictures of the old plantation types'. The book was part of Dutton's catalogue of juvenile literature and claimed that it would 'appeal to the little folks, whether they have been in Dixie land or not'. As was often the case, many consumers were in the latter category.[27]

National advertisements employing Old South imagery were not limited to black stereotypes alone. Gorham, a company that manufactured table silver, advertised in *Life* in 1928 that its Colfax line of silverware was 'a delightful reflection of southern hospitality' much like that used by the 'charming hostesses of the Colonial South'. A Philadelphia distillery advertisement campaign for 'Dixie Belle Gin' promoted an image of upper-class leisure often associated with the planter South. J. P. Lippincott, a book publisher also located in Philadelphia, carried a large inventory of romance novels in which a Northern man married a Southern woman, the literary symbol of sectional reconciliation.[28]

Life magazine included a number of national advertisements for southern-based companies. While Coca-Cola was the best-known manufacturer based in the South during this period, the company was committed to being *national* in its appeal, even though its chairman, Asa Candler, was a staunch Confederate. Other southern companies, however, participated in projecting the Old South, Dixie brand. The Natchez Baking Company of Mississippi advertised its 'Ole Missus Fruit Cake' in 1924 as 'made in Dixie'. Nunnally's, a candy manufacturer based in Atlanta, advertised its box of chocolates as being renowned 'ever since the old days of the storied South'. Using the image of a southern belle being wooed by her beau, the candy was described as having 'subtle undertones of flavor that emphasize so delightfully its Southern witchery', well before Margaret Mitchell introduced the world to Scarlett O'Hara.[29]

An important advancement in advertising in the 1920s and 1930s was what historian Roland Marchand calls the 'social tableaux'. Advertisements of this type usually depict a 'slice of life' serving as reflections on society. While it was not the intention of advertisers – profit was the motive – these tableaux do depict scenes that reflect contemporary cultural values. The advertisement's message is not simply a slogan, but an entire story created by the scene.[30]

One example of the social tableau that perpetuated the Dixie/Old South theme was an advertisement for Crab Orchard whiskey, distilled in Louisville, Kentucky. The 'slice of life' is a scene of the old 'Crab Orchard Springs Hotel' where people came for such 'Southern delicacies as barbequed squirrel' or 'roast 'possum and candied yams'. They washed it down, of course, with bourbon whiskey, 'a flavor which even the flower of old-time Kentucky's gentility praised'. Crab Orchard Whiskey, in this case, was branded as 'old-fashioned' Dixie with the accompanying value of being associated with southern gentility, and became, according to the claim, 'America's fastest-selling whiskey'.[31]

Even as advertising represented a modern trend, when it came to buying Dixie, it meant selling tradition. Southern consumers purchased objects that then became artefacts of their belief system – in this case, the Lost Cause. Advertisers did not necessarily have to create a market for buying items linked to Confederate memory to play on consumers' anxiety about a changing South, because it already existed. Yet they could perpetuate that anxiety in the way they sold their products by suggesting, for example, that if a particular book was not purchased, then the story would be lost. Like national companies, those who advertised in the *Confederate Veteran* offered products that they suggested readers needed or could not live without. In doing so, they helped perpetuate the Lost Cause through mass consumerism. Whether the advertising was martial in spirit, as it was for southern companies, or focused on the sentimental and exotic, as with the advertisements of northern companies, both regions help illustrate the profitability of branding Dixie (that is, the South) in a mass consumer society. 'Dixie' became a brand in a broad sense, and as a brand it conveyed the values of the product as well as the values of white society.

The most obvious cultural expressions of the Lost Cause are the monuments that dot the landscape of the South, and has been discussed in recent works including, *Monuments to the Lost Cause: Women, Art, and the Landscape of Southern Memory*[32] and my own work on the United Daughters of the Confederacy. Yet the landscape hosted other manifestations of Confederate culture, most notably history museums either founded for the purpose of housing Confederate relics and thus becoming Confederate museums or were southern state museums of history founded with Civil War artefacts collected by Confederate organizations. The cities of New Orleans and Richmond, both established Confederate museums, in 1891 and 1896 respectively.

These and other institutions served as the repositories and exhibit spaces of Confederate culture. The Confederate Memorial Literary Society, a group of women who also held membership in the UDC, founded the Confederate Museum in Richmond. Housed in the former White House of the Confederacy, which had been home to Jefferson Davis and his family during the Civil War, the museum exhibited and housed artefacts from the former states of the Confederacy. As one writer to the *Confederate Veteran* put it, southerners owed it to the 'memories of its Confederate dead' to preserve their state's 'best relics'. The Daughters agreed, and the material culture they gathered provided a three-dimensional narrative of the war.[33]

Around the same time, again in the city of Richmond, the Confederate Memorial Institute, known to contemporaries as the 'Battle Abbey of the South', was conceived. In 1895 Charles Broadway Rouss offered the UCV $100,000 to construct the memorial to house a portrait gallery of 'military heroes and civil leaders of the Confederacy'. Public debate and internal conflicts delayed construction of the building, but in 1921 it opened to the public and soon became another of the city's Confederate landmarks. Yet, unlike monuments, the Battle Abbey helped perpetuate Confederate culture among a new generation of southern children. Just as they studied Confederate heroes in their public schools, children from around Virginia were chaperoned by their teachers on field trips to the Confederate Memorial Institute. In 1922, 5000 school children visited the museum to see portraits of Robert E. Lee and J. E. B. Stuart – Confederate heroes they studied in their classrooms.

The Battle Abbey proved to be a tourist attraction as well. Throughout the 1920s, museum averaged 10,000–12,000 visitors per year – many of them from outside of Virginia. Several travelled from Northern states, especially New York. This trend continued into the 1930s, and in the midst of the Great Depression the state appropriated funds to keep the Confederate Memorial Institute open. What is apparent through a study of the Institute's records is that by the mid-1930s, Richmond had developed a tourist industry around Confederate culture. A board member of the Institute wrote to the city's Director of Public Safety asking that the city policemen be provided small cards that showed the location of all of Richmond's Confederate sites noting that this 'may help in a small way to get more people to visit the Battle Abbey and Confederate Museum'. The public safety director responded positively: 'I shall be very glad indeed to have our policemen carry such cards as you may care to print with the information as to the points of interest around the city.'[34] And indeed, they did.

Sounds of Dixie

Perhaps the most entertaining and effective way to brand Dixie was through popular culture, especially the sale of sheet music about the region. Before there was radio, there was parlour music, and in the years between 1890 and 1930, sheet music on the theme of 'Dixie' was as popular as it gets. In the 1890s, the genre of music on the South known as 'coon songs' were popularized, but by 1910 they had been replaced by the music of 'Tin Pan Alley' with more of a focus on the South generally, than specifically on southern blacks, although racial stereotypes continued to exist in both lyric and in the art work that appeared on sheet music covers. Interestingly, this music was not being produced in the South or by southern musicians, but in the North.

New York City's 'Tin Pan Alley' musicians and songwriters wrote hundreds of songs about the South, often using 'Dixie' in their titles. Music publishers in New York, Detroit, and Chicago sold thousands of copies of these piano pieces, which were played at home and made popular by Vaudeville acts like the Misses Campbell who popularized the song 'You're as dear to me as "Dixie" was to Lee'.[35]

Ironically, as well as most interesting, Jewish immigrants living in New York City and who had never laid eyes on the American South, wrote most of these Dixie songs. Irving Berlin wrote several pieces on Dixie including, 'The Dixie Volunteers' (1917) and 'When It's Night Time Down in Dixieland' (1914). His publishing company, Waterson, Berlin & Snyder, produced several Dixie pieces. Jerome H. Remick, who had publishing offices in Detroit as well as New York, was responsible for a substantial part of the market for sheet music on Dixie with titles such as 'Down South Everybody's Happy' (1917).

'Are you from Dixie?', 'How's Ev'ry Little Thing in Dixie?', 'When the Sun Goes Down in Dixie', and 'She's Dixie All the Time' touted the leisurely pace of life in the South as the antithesis to modernity, as songwriters told of days gone by, or of a life they wished they enjoyed, or perhaps, missed. They reveal a fascination with the 'exotic' South, a place in a warm climate with happy-go-lucky blacks, of genteel women, of 'moonlight and magnolias'. The lyrics from 'Dixie Lullaby' published in 1917 by a Chicago music publisher are typical:

> There's a tale that they tell about Dixie
> It's heaven on earth so they say,
> With the birds and the flowers where I spent happy hours
> And the tho't of it takes me away

CHORUS
Down in Dear old Dixie where the flowers bloom
Down in Dear old Dixie in the month of June
Floating down the river in a birch canoe
And singing love's song to you
In the fields of cotton where I used to roam
On the old plantation of my Southern home,
Back in Alabamy, back beside my mammy
That's my Dixie Lullaby . . .

Isaac Goldberg, who lectured and wrote about American popular music, published his treatment of the genre in his book entitled *Tin Pan Alley* (1930). In it, he relates the story of the famous lyricist Gus Kahn who was asked why the 'song boys' of the North all wrote about the South. Kahn replied that 'Southern place-and-State names lent themselves to rhyming'. Goldberg added what he believed to be a 'deeper reason' for why northern songwriters were fascinated by the South. 'Paradise is never where we are. The South has become our Never-never Land – the symbol of the Land where the lotus blooms and dreams come true.'[36]

For Goldberg, the South was an exotic place. And for songwriters as well as advertisement men who sought to brand the region, the South represented a respite from the afflictions of modernity. If, for northerners, 'paradise is never where [they] are', then 'Dixie' was paradise, even if the reality did not match the myth. They needed only the idea of Dixie to make it seem real, and the nation's consumers ate it with a spoon.

What can be concluded from these expressions of Confederate culture in the era of the 'New South'? What are the implications for all members of a society permeated by Confederate culture and Lost Cause ideology? In his book *The Culture of Defeat*, Wolfgang Schivelbusch sees in the Lost Cause a 'program for the conservation of national identity', in this case, an identity of southern nationalism, which at its core embraces and seeks to defend white supremacy.[37] That being the case, then the most obvious implications of Confederate culture were for race relations in the South in the late nineteenth and for most of the twentieth centuries. Celebrating a culture of white supremacy provided vital reinforcement to *de jure* and *de facto* segregation.

Pro-southern literature, Confederate consumer culture, museums and music dedicated to the preservation of that culture were important not only for assuaging white guilt over Confederate defeat, but because together they helped preserve conservative traditions tied to Confederate nationalism. It should come as no surprise that preservers of Confederate culture looked to recreate the past by preserving traditional gender relations, and were as a group (men and women) opposed to women's suffrage. Additionally, they regarded race relations in the days of slavery, of black deference to white masters, as instructive for their own times. Younger generations of blacks, they believed, needed to revive those traditions of the past by being deferential domestic servants and labourers. The long-term implications of Confederate culture and Lost Cause mythology were, most importantly, its impact on both race relations and the South's relationship with the nation – serving as it did to limit racial progress and alter sectional reconciliation on terms acceptable to the South.

Confederate defeat in historical memory, therefore, was remembered as a loss, but in strictly military terms. Otherwise, the reality of defeat was swiftly replaced with the myths of the Lost Cause. Commemorations and celebrations of heroic deeds and of regional values related to states' rights and white supremacy took the place of mourning and repentance. Defeat was recast as the victory of values and manifested itself in both southern popular culture and American popular culture. The battle flag became a recognizable icon of these values in the former Confederacy, while 'moonlight and magnolias' was associated with the South nationally. In the end, Confederate defeat demoralized the white South only in the short term; over time, it was replaced by a celebration of honour and heritage, a celebration that continues in some circles to the present day.

Notes

1. Charles Reagan Wilson, *Baptized in Blood: The Religion of the Lost Cause, 1865–1920* (Athens: University of Georgia Press, 1980); Gaines M. Foster, *Ghosts of the Confederacy: Defeat, the Lost Cause, and the Emergence of the New South, 1865–1913* (New York: Oxford University Press, 1987). More recent examinations of the Lost Cause include Karen L. Cox, *Dixie's Daughters: The United Daughters of the Confederacy and the Preservation of Confederate Culture* (Gainesville: University Press of Florida, 2003); William Blair, *Cities of the Dead: Contesting the Memory of the Civil War in the South, 1865–1914* (Chapel Hill: UNC Press, 2004).
2. David Blight, *Race and Reunion: The Civil War in American Memory* (Cambridge: Belknap Press, 2001); and Nina Silber, *The Romance of Reunion: Northerners and the South, 1865–1900* (Chapel Hill: UNC Press, 1995).
3. Richard S. Tedlow, *New and Improved: The Story of Mass Marketing in America* (New York: Basic Books, Inc. 1990), p. 14; Stephen R. Fox, *The Mirror Makers: A History of American Advertising and Its Creators* (New York: William Morrow

and Company, Inc., 1982), p. 22. See also Jib Fowles, *Advertising and Popular Culture* (Thousand Oaks, CA: Sage Publications, 1996), pp. 32–3. Other secondary sources on the history of advertising in the United States in the late nineteenth and early twentieth centuries include Frank Presbrey, *The History and Development of Advertising* (New York: Doubleday, Doran & Company, 1929) and James D. Norris, *Advertising and the Transformation of American Society, 1865–1920* (New York: Greenwood Press, 1990).

4. On Confederate holidays, see Karen L. Cox, 'Confederate Holidays' in Len Travers ed., *American Holidays and National Days: An Encyclopedia* (Westport, CT: Greenwood Press, 2006); on the pro-Southern textbook campaigns, see Fred A. Bailey, 'Free Speech and the Lost Cause in the Old Dominion', *Virginia Magazine of History and Biography*, Fall 1995.
5. Paul Gaston, *The New South Creed: A Study in Southern Mythmaking* (New York: Alfred Knopf, 1970).
6. A bibliography of Thomas Nelson Page's works in the genre of the Lost Cause can be found online at: http://bindings.lib.ua.edu/gallery/nelson_page.html#bib. See also Joel Chandler Harris, *Uncle Remus: Legends of the Old Plantation* (1881).
7. Cox, *Dixie's Daughters*, pp. 93–117.
8. Ephemera (ca. 1895), B. F. Johnson Papers, Virginia Historical Society (VHS).
9. Ibid.
10. T. J. Jackson Lears, 'From Salvation to Self-Realization: Advertising and the Therapeutic Roots of Consumer Culture, 1880–1930' in Richard Wightman Fox and T. J. Jackson Lears eds, *The Culture of Consumption: Critical Essays in American History, 1880–1980* (New York: Pantheon Books, 1983), p. 7. See also Roland Marchand, *Advertising and the American Dream: Making Way for Modernity, 1920–1940* (Berkeley: University of California Press, 1985).
11. Michael Walzer in Fox and Lears eds, *The Culture of Consumption.*
12. Norris, *Advertising and the Transformation of American Society*, p. 29.
13. Ibid., pp. 31, 33; Fox, *Mirror-Makers*, pp. 28–35.
14. Tedlow, *New and Improved*, p. 14.
15. Ibid. pp. 14–15; Norris, *Advertising and the Transformation of American Society*, p. 99.
16. Russell W. Belk and Richard W. Pollay, 'Images of Ourselves: The Good Life in Twentieth Century Advertising', *Journal of Consumer Research*, vol. 11 (March 1985), 888.
17. Grace Hale, *Making Whiteness: The Culture of Segregation in the South, 1890–1930* (New York: Pantheon Books, 1998), p. 146. Hales's argument about buying and selling within the region as well as outside the region is useful; her analysis of advertising focuses on black stereotypes in national advertising.
18. On the use of blacks in advertising, see Marilyn Kern-Foxworth, *Aunt Jemima, Uncle Ben, and Rastus: Blacks in Advertising, Yesterday, Today, and Tomorrow* (Westport, CT: Greenwood Press, 1994); M. M. Manring, *Slave in a Box: The Strange Career of Aunt Jemima* (Charlottesville, VA: University Press of Virginia, 1998), p. 9. See also Hale, *Making Whiteness*, pp. 121–98.
19. The first run of the *Confederate Veteran* was between 1893 and 1932. It was the official organ of all Confederate heritage societies. Hereafter cited as *CV*.
20. 'Tichenor's Antiseptic Refrigerant for Wounds', advertisement, *CV* (April 1906).

21. 'Stars and Bars', Advertisement, *CV* (March 1906); Allen & Ginter Trading Cards, Benjamin Meade Everard Scrapbook, 1889–90, VHS *Souvenir Radio Log WRVA, Richmond Virginia 'Down Where the South Begins', The Edgeworth Tobacco Station* (Richmond, VA: Larus & Bro. Co., 1929), VHS.
22. 'S. Thomas & Bro.' Advertisement, *CV* (1910); 'S. N. Meyer' Advertisement, *CV* (April 1906); 'The Game of Confederate Heroes', *CV* (July 1898).
23. The Dixie Flyer and the Cotton Belt Route advertised regularly in the *Confederate Veteran* between 1900 and 1920.
24. M. C. Lilley and Pettibone advertised in the *CV* between 1900 and 1920; 'The Couch Beautiful', Advertisement, *CV* (1913); 'Dixie Artificial Limb Company', *CV* (February 1906).
25. Advertisements for the McNeel Marble Company and the Muldoon Monument Company appeared consistently in the *CV* between 1910 and 1920.
26. Kern-Foxworth, *Aunt Jemima, Uncle Ben, and Rastus*, pp. 29–41; Hale, *Making Whiteness*, pp. 151–68; and Manring, *Slave in a Box*, pp. 79–100.
27. 'ABC in Dixie', Advertisement, *The Critic* (December 1904).
28. 'Gorham Silver', Advertisement, *Life* (5 October 1928), 92; 'J. P. Lippincott Holiday Books', Advertisement.
29. 'Ole Missus Fruitcake' and 'Nunnally's' advertisements in *Life* (November 1924; August 1920).
30. Marchand, *Advertising the American Dream*, 164–8.
31. 'Crab Orchard Whiskey', advertisement, *Life* (1 June 1935), 8.
32. Cynthia Mills and Pamela H. Simpson (eds), *Monuments to the Lost Cause: Women, Art, and the Landscape of Southern Memory* (Knoxville: University of Tennessee Press, 2003).
33. Cox, *Dixie's Daughters*, pp. 98–9.
34. S. L. Carter to Miss Irene Harris, Battle Abbey, 19 February 1938 and John A. Cutchins to S. L. Carter, 18 February 1938; Board of Lady Managers Correspondence, Confederate Memorial Institute Collection, Virginia Historical Society, Richmond, Virginia.
35. Isaac Goldberg, *Tin Pan Alley: A Chronicle of the American Popular Music Racket*, (New York: The John Day Company, 1930). 'You're as Dear to Me as "Dixie" Was to Lee' (New York: Leo Feist, Inc., 1917).
36. Goldberg, *Tin Pan Alley*, pp. 45–6.
37. Wolfgang Schivelbusch, *The Culture of Defeat: On National Trauma, Mourning, and Recovery* (New York: Metropolitan Books, 2003), p. 59.

12

Gallipoli to Golgotha: Remembering the Internment of the Russian White Army at Gallipoli, 1920–3

Anatol Shmelev

On 17 November 1920 the final remnants of the Russian army evacuated the Crimea in the face of the advancing Red army. Over the course of the preceding week, nearly 150,000 people had been loaded onto 126 ships of the line, troop transports, tugboats, barges, yachts – anything that could float – and sent west, to the Bosphorous. Not everyone could be evacuated, and the departing troops and civilians bore witness to tragic scenes of officers shooting themselves on the quays as they watched the overloaded ships cast off.[1] The Civil War had been lost, and the Russian army – defeated, demoralized, homeless and unwanted – appeared at the gates of Europe, awaiting its further fate.

Adding insult to injury, because many ships left port without adequate supplies in order to accommodate more people, passengers on arrival in Constantinople were hungry and thirsty, and were forced to sell their wedding bands and baptismal crosses for food and water. Some ships had spent several days at sea with no fresh water. First off the ships were the wounded and sick, as well as civilian refugees. The officers and soldiers of the army remained aboard, but in less confined circumstances and with adequate food and water. Immediately prior to the evacuation, the Russian command had negotiated terms with the French, who had taken responsibility for the fate of the evacuees against the collateral of the ships of the Russian fleet.[2] The question remained of where to send them.

The Russians looked upon their arrival at the gates of the Bosphorous with irony: for centuries it had been the dream of the Empire to retake 'Tsargrad' – 'Tsar-City' as they still called Constantinople (the name Istanbul was never used) – and finally the Russian army had arrived, not as conquerors, but rather as a vanquished and homeless armed mass.[3]

The question of what to do with troops was soon resolved: regular infantry, cavalry and artillery units, as well as military colleges, technical and engineering units combined into I Army Corps, were to proceed to Gallipoli, while the Cossacks were interned at Çatalca (Chataldja) and on the island of Lemnos. By 19 December 1920, 26,596 people were counted to have descended on the sleepy little town of Gallipoli and the army encampment – soon named the Valley of Death – located a few miles further down the peninsula. Of these, 23,518 were officers and soldiers, the rest wives and children.[4] An additional 24,843 Don and Kuban Cossacks were interned at Çatalca and on the island of Lemnos respectively.

It immediately became clear that the town of Gallipoli was unprepared to house such a large number of internees: many buildings were still in a state of disrepair from the battles of 1915 and the earthquake of 1912. Nor were there enough tents in the camp to give shelter to every new echelon disgorged from the arriving ships: many had to remain on board or sleep under the open skies. Entire units had no shoes, clothes were in tatters; insufficient rations, cooking facilities, latrines and sanitary equipment immediately began to take their toll on the new arrivals with the spread of typhus and other diseases. Some spent their entire stay in Gallipoli in holes dug in the ground and covered by a tarp or thatched roof.[5] Small wonder that the name Gallipoli was soon converted by the Russians into *Goloe pole* – 'Naked field', settled by an apathetic, tired, demoralized and broken mob.

The main problem was one of morale, and though the factors referred to above undoubtedly exerted a negative influence, it was the completely uncertain situation of the army that caused a number of desertions in the first days. Many of these were not necessarily the worst elements: some had simply lost faith that they would be sent to continue the struggle on the Polish front or in the Far East. For this reason, the first steps undertaken by the command were the establishment of a bulletin to combat rumourmongering and the re-establishment of firm discipline.

The commander of I Corps was Lieutenant-General Alexander Pavlovich Kutepov, formerly a Guards officer and one of the closest associates of Commander-in-Chief General Petr N. Wrangel. Kutepov immediately set about imposing discipline on the demoralized troops, many of whom had either never been exposed to barracks-style discipline or had forgotten it after nearly seven years on the battlefield. Kutepov gave no quarter to transgressors, applying penalties ranging from arrest for varying periods of time, up to a firing squad.[6]

One of Kutepov's early measures was to explain and instil in the troops that 'We remain the same soldiers that we have been up to now; the appellation "refugees" used by the French to describe us, is a fiction and should not reflect anything.'[7] He insisted that all officers and men should be properly dressed in their uniforms, should salute and should perform all the elements of military ritual; the use of foul language was prohibited.

Many felt he was too strict, even cruel and brutal in his punishment. However, he was fully supported by Wrangel in his efforts. Indeed, in December 1920 and January 1921, Wrangel issued orders reinstating courts martial, officers' courts of honour and even duelling as measures to increase morale and halt the disintegration of the concept of honour.[8] Kutepov's measures led to the desired effect. Many Gallipoli internees surprised themselves when they suddenly saw the change in their own morale, for there was a key moment in March 1921, when the meaning of Kutepov's measures suddenly became obvious.

Relations with the French – never good – had deteriorated considerably in the spring of 1921. There was a small French garrison in the town, which, while it was not actually guarding the Russians, often came into conflict with the latter over questions of authority and maintenance of order. As early as December, the French attempted to confiscate Russian arms, to which the Russian command responded with a secret order to create hidden weapons depots in the units. This was followed by recruitment campaigns for the French Foreign Legion, settlement in Brazil and even repatriation to Soviet Russia. There was reason to believe that the French were planning to cease victualling the army on the first of April 1921. In order to forestall such an attempt to starve the Russians into submission and dispersal, the high command began to prepare plans for a march on and the subsequent occupation of Constantinople, the purpose of which was to draw attention to the army's situation. These plans were kept in secret, and the preparations – drills, training and manoeuvres – were presented to both the troops and the Allies as necessary for a possible overland trek to the Balkans.[9]

Nevertheless word spread of the real purpose, so that one day, when Kutepov was called to Constantinople for a meeting with the French commander of the occupation forces, and rumours spread that he was being detained against his will to keep him from returning to the army, the officers and soldiers were convinced that the march on Constantinople was at hand. In their later memoirs, officers noted that suddenly their antipathy towards Kutepov dissipated and was replaced by an attachment so strong that when Kutepov finally did return to Gallipoli, he was physically carried by his men above their heads from the ship to his headquarters in the town, and everywhere greeted with overwhelming hurrahs.[10] In his absence, General A. V. Turkul, commander of the Drozdovsky Division, had organized drills and parades (with arms) in the camp, which had the effect of causing the French garrison (500 Senegalese riflemen with 28 machine guns) to surround their own barracks with barbed wire.[11]

From this point, Kutepov's position was established, and the crux in the transfiguration of the army – and of the entire Gallipoli experience – was reached. Given that Kutepov could now speak with the French command as commander of an army, rather than a mere refugee, the French were forced to treat him – and the army – with respect, which helped to preserve not only the physical military structure but also the lives of the individuals who composed it: French plans to disperse the troops, repatriating most of them to death or servitude in Soviet Russia, were defeated.[12]

Under Kutepov's leadership, Gallipoli began to evolve from a place on a geographical map into an idea. Originally, this idea was to serve a very specific need: preparation for the universally expected (among the Whites) 'Spring Campaign', that is, the expectation that the spring of 1921 would bring a wave of anti-Soviet uprisings in Russia that would sweep away the Bolsheviks. The Russian Army was to provide the cadre for the military victory in this campaign. As spring approached, a rebellion broke out at the Kronstadt naval base, which raised hopes that the hour had come. But this rebellion was quickly suppressed, leaving the Whites with no choice but to prepare for an undetermined future. Wrangel and his associates negotiated feverishly with the Balkan countries to allow the troops interned on the Gallipoli peninsula and Lemnos to settle there, and finally agreements were reached with Bulgaria and Serbia that allowed the bulk of the army to find new homes in those countries. While this was an enormous victory for Wrangel and the troops, the question that was now posed to them was how to maintain 'in semi-secret fashion' the cohesion of an army, so that it would be able to fulfil its duty when called upon to liberate Russia.[13]

Gallipoli was perceived as a victory of the spirit both during its time and, especially, afterwards. Among émigré Russian officers, the name of the peninsula became synonymous with dignity, honour and patriotism. The humiliations of defeat at the hands of the Reds, of being treated by French colonial troops as though they were under guard, of the haughtiness of French officers, of inadequate rations, of harsh conditions became part of the narrative of overcoming all odds to survive, not only individually but as an army. Defeat at the hands of the Reds was forgotten in the course of the new struggle against the French, against demoralization and against émigré defeatism.

The strength of morale at Gallipoli and the way it was maintained through firm discipline can be illustrated by the response to dissent. For example, much of the left-leaning émigré press, from the Socialist-Revolutionary *Volia Rossii* to the Constitutional-Democratic *Posliedniia novosti*, were positively venomous towards the Gallipoli internees, whom they considered either reactionaries or willing dupes of reactionaries, rightly punished by history. Kutepov and Wrangel were singled out in this criticism. Yet these newspapers were not only not prohibited but hung out on public display and discussed in the tents and at 'verbal journals' (that is, meetings with lectures).[14] At the height of the émigré press campaign against Wrangel and Kutepov, the latter decided to show that no one was being held in Gallipoli against his will. On 23 May 1921, Kutepov issued an order giving everyone who wished to leave the ranks of the army three days to do so. Some 2057 men (approximately 10 per cent of the total troops interned on Gallipoli) chose to leave; the rest closed ranks. Thereafter, Kutepov's discipline was indeed severe: Colonel P. N. Shcheglov, tried by a military court martial for propaganda in favour of dispersing the Russian army, was executed by firing squad on 30 June 1921.[15] After having given everyone 'weak in spirit' ample time to leave the ranks of the army and join the refugee camp, the army leadership felt justified in demanding full loyalty from those who remained.

Another form of response to the attacks of the left wing of the political emigration came from conservative authors such as future Nobel Prize laureate Ivan Bunin, Ivan Lukash, Ilya Surguchev, Petr Krasnov, Nikolai Breshko-Breshkovskii, Aleksandr Kuprin and others, many of whom were invited to visit the camp. Their impressions of Gallipoli and its significance were collected and published as *Zhivym i gordym* (Belgrade: Izdanie Obshchestva Gallipoliitsev, 1923). Writing in *Dni*, a left liberal daily edited by Alexander Kerensky, Boris Mirskii reviewed *Zhivym i gordym*. He was appalled that the Gallipoli internment could be placed on a par with the most glorious pages of Russian military history, and concluded his criticism with the hope that someday 'both Dzerzhinskii's *chekists* [early Soviet secret police, renowned for their brutality] and Kutepov's hangmen would be grouped together in one bloody category'.[16]

Not only visitors but also the internees at Gallipoli themselves responded to attacks from the left, sometimes concurrently taking a swipe at the saccharine sweet tones of the literary figures who defended them: 'we're all sick of marching and saluting', declared one internee, and 'we don't want to be remembered' by amateurish double-headed eagles laid out of rocks and seashells before regimental guard-posts. He compared the glowing reports of camp life by sympathetic visitors to descriptions of peasant life in Russia by urban summer vacationers: 'sweet and mistaken'. This anonymous author of a letter to *Obshchee dielo*, one of the newspapers most loyal to General Wrangel (though edited by a former socialist), described a widening gulf between the senior officers and subalterns and men that had been unnoticeable earlier, on the battlefield, but was growing intolerable in Gallipoli, where Kutepov's emphasis on discipline was serving to bring out the martinet in many senior officers. And yet, the author noted, he and his fellow officers were willing to accept this and more as long as they were motivated by the conviction that the army was necessary to free Russia from the Bolsheviks, and that the only way to preserve the army was to fully submit to General Wrangel.[17]

Through the course of events, the defeat at the hands of the Bolsheviks was turning into a victory over the attacks of émigré politicians and political groups, over the effort of the French to disband the army and over the indifference or coldness of the rest of Europe. 'The year of its exile became the year of its decisive victory', declared Ivan Lukash, 'the army achieved a victory of the spirit.'[18]

One method of improving morale was by creating cultural activities that would put the internees' spare time to some uplifting purpose. Despite meagre resources, and inhospitable and bleak surroundings, many officers made use of their pre-war professions: teaching, science, journalism, while others relied on inborn talents. A number of publications began to circulate, with a variety of content. Most of these were printed in very limited runs, very often a single copy of each issue, for the officers of a single unit, but sometimes even these received broad renown. One example was *Veselye bomby* (Happy Bombs), the journal of the third Drozdovsky Artillery Battery, which found among its ranks two very clever cartoonists. One of the caricatures, entitled 'The Final Match', made their tents quite popular for a time. It depicted a football field with two goals, labelled Serbia and Brazil. One team of Senegalese was aggressively pushing the ball towards Brazil, and indeed, the referee (depicted as Trotsky) had already raised the whistle to his lips to signal a goal, when suddenly General Kutepov jumped high and, striking the ball with his head, sent it flying towards Serbia. 'Both in its idea and its artistic qualities the caricature was very successful, and soldiers and officers came from all over to our battery to ask to see it. At a general meeting of teachers on 30 May [1921], Prof. V. Kh. Davatts, underscoring the meaning of this caricature, gave a special lecture on the troubling events of these days.'[19]

Another aspect of the recognition of the need to find proper outlets for the energy of the internees and ensure that – no matter what the future held – they would be in a capacity to use their knowledge and experience for the benefit of themselves, each other and Russia was the army command's decision to set up an education system involving lectures, courses, examinations and certification. A complete lack of study and teaching materials, handbooks and other pedagogical accoutrements was a major handicap, although the army could provide a large number of qualified teachers in a variety of fields of knowledge, up to and including former university professors, such as the aforementioned V. Kh. Davatts, who had in civilian life been a professor of physics at Kharkov University.

Of course, much of the educational programme concentrated on military science, but a good deal of the knowledge required, especially for artillery and engineering officers, could be translated into civilian life as well: mathematics, physics, chemistry and so on. Five military colleges were set up in Gallipoli, alongside three officers' training courses and numerous other military and civilian courses, in addition to a school for children in the town.

Sports, especially football, were organized. Chess matches took place, with both boards and pieces handcarved from the wood of local trees and bushes. The same material served for the erection of churches and iconostases, with tin cans artfully recrafted into candleholders and chandeliers. Choirs, both church and secular, were formed; literary and artistic clubs were established, and even two theatres, each with its own company of actors, were created. By the summer, a library had been established, and regular exhibitions of arts and crafts attracted both locals and visitors from the town. 'On the shores of the Dardanelles, General Kutepov created a microcosm of Russia, and every participant in this astounding phenomenon felt himself not a passive [onlooker], but an active creator of new values.'[20] It was this feeling that made Gallipoli stand out from the other numerous internee camps in Poland, Estonia and China, some of which held many thousands of Russians and existed for many years (particularly in Poland), but never came close to producing a collective memory or spirit on a par with Gallipoli. Even Lemnos and Çatalca, geographically the closest camps and related to Gallipoli by derivation from a common source and similarity of circumstances, were pale shadows of the latter.

Witnesses of the conditions on Lemnos and in Gallipoli reported starkly contrasting situations. Physically, both groups of Russians were in roughly similar conditions: extremes of weather;[21] lack of food and water; inadequate shelter, clothing and blankets. Yet the morale on Lemnos was substantially worse: the officers and men were dirtier; less disciplined; more inclined to complain, to disobey orders, not to salute and to desert to the refugee camp, from where they hoped to escape further to Greece, Serbia or Bulgaria. Others signed up for the French Foreign Legion, and a small number even deserted to become bandits in the hills of the island. When, in June, the ship *Reshid-Pasha* repatriated a group of White Army veterans to Soviet Russia, it carried 475 Gallipoli internees willing to return. These were joined by over 2500 Lemnos internees: an indication of the difference in the spirit of both camps.[22] This difference is also quite noticeable in the later publications (memoirs, anniversary almanacs, brochures and periodicals) of participants: those on Lemnos, Çatalca, Bizerta (Tunisia), and other camps left very little record, and that often marked by tones that did not appear in Gallipoli publications. Thus, in the preface to his memoirs of Lemnos, S. Rytchenkov writes that his book is 'a page in the tragedy of the Russian people'.[23] One is hard put to find any similar type of admission coming from a Gallipoli veteran: to them Gallipoli is a page of glory.

One year and one day following the first appearance of Russian troops on Gallipoli, ships arrived to take the members of the army to their new homes: Bulgaria and Serbia had offered to accept almost the entire population of the Gallipoli and Lemnos camps for settlement. Most of I Army Corps was to go to Bulgaria, while the cavalry and Cossacks were primarily sent to Serbia. On 14 December, General Kutepov embarked with the final column of refugees, and the Russians bid farewell to the peninsula whose name had become synonymous with a patriotic rebirth that was to last for many of them until death.[24]

With their resettlement, the officers and soldiers took with them a keen understanding of what had happened in Gallipoli: a defeated mob had been transformed into a disciplined and tightly bound fighting force. The main concern now was to retain this camaraderie and discipline in the expectation of a new stand against the Bolsheviks; as Kutepov put it, 'If we have discipline, we will have an Army; if we have an Army – we will have Russia.'[25] In fact, the creation of a disciplined fighting force at Gallipoli was seen as the antipode of the Russian Imperial Army's dissolution in 1917,[26] and this view made the veterans all the more conscious of their participation in a process far more important than could be divined from a superficial examination of their conditions. The theme of transformation was almost universal in later reminiscences, although the degree and nature of this transformation varied according to the background and purposes of the memoirist. One of the elements this transformation touched was the Whites' image of the meaning and goals of their own struggle. Vladimir Dushkin put this succinctly: 'In Gallipoli we lost and buried the real White Idea, but we acquired the even more spiritual, unreachable, clarified, shining [prosvetlennuiu] White Dream. I think we all understood this, but could not or would not allow ourselves to admit this. It seemed treasonable.'[27] Terse as this statement is, it is pregnant with a meaning that can only be understood by a participant (the more so as Dushkin offers no explanation).

Of course, no spiritual transformation lends itself to explanation in words, the more so to succeeding generations. Yet, though the full significance of any experience can never be felt by a non-participant, the reality and effect of the phenomenon can be to some degree described. The Gallipoli Society that evolved from the internment had more than an average amount of social glue bonding its members together: rarely do defeated armies of exiles expend so much energy and resources on publishing activity, social events, erection of monuments, purchases of real property and cemetery sections as did the Gallipoli veterans, especially considering the depth of their poverty and the hardship of their working and living conditions.

Some aspects of the transformation have already been touched upon: the change from a demoralized, defeated mob to a disciplined army; the creativity of expression available in camp publications, theatre, choirs, artistic studios; the sense of national pride strengthened by the small, but very keenly felt, victories over Allied attempts to disarm and disband the Russians. But another important quality, underscored by Dushkin in his cryptic manner, also contributed heavily to the transformation.

In their defeat, the Whites struggled with the question *why?* It had been clear to them during the struggle that they were in the right, and yet they had been forced into exile. This led to other questions: Was the Red victory only temporary? Did the Whites do wrong to engage in combat, and did this ultimately lead to a strengthening of the Bolshevik regime? These questions were under constant discussion on Gallipoli, as, to be sure, they were in Poland, China, Estonia, and every other country in which the Whites found themselves. The questions were magnified by the attention devoted to Gallipoli by the leftist émigré press, which maintained that the entire White Movement was a mistake and a crime. Those who could find no answer to such questions or who concluded that the struggle was mistaken typically left the army, either to return to Soviet Russia or to disappear in the refugee mass. But for those who remained in Gallipoli, painfully and slowly, the questions were answered, and a new understanding of the experience of civil war and defeat emerged. 'Will the White Army be needed in the future, were our leaders correct in calling us to battle, did we not commit a crime in struggling against the people[?] These damned questions tortured us for months. [. . .] Who are we: criminals or unrecognized friends of the people?', asked S. M. Shevliakov at a session of his 'verbal journal' on 25 October 1921. His answer was unambiguous: 'We know now that blood was spilt in the name of peace for Russia and popular well-being. After Kronstadt we understood that the people on their own had not the strength to overthrow the Bolsheviks.'[28]

This, perhaps, clarifies somewhat Dushkin's cryptic commentary on the transformation. The abortive rebellion at Kronstadt in March 1921 illustrated to the Gallipoli internees that armed struggle was the only method of ridding the country of the Bolsheviks, thus justifying the efforts of the previous years and reinforcing the insistence of the army command that the only possible path to victory over the Bolsheviks was in complete dedication to Russia and tight unity. Even as the reality of continuing the struggle receded with every day and year in emigration, these abstract concepts – the White Dream – held fast.

Two symbols created during the internment served to remind the veterans of their time on the peninsula: the Gallipoli cross and the Gallipoli monument. The cross is one of the most common physical forms of an order or medal, but in the case of the Gallipoli cross, the form was further augmented by the substance of how Gallipoli was defined in terms of the Passion of Christ. The order establishing the cross included the following text: 'You bore your cross for an entire year. Now, in memory of the Gallipoli internment, you will bear this cross on your chest. Unite the Russian people about this cross.' In the minds of its internees, Gallipoli assumed the character of a place of suffering, perseverance and penance, whose purpose was, as one veteran put it, to cleanse the soul: 'Strong yet is our sin, the essence of which is a weakness in faith in general, of faith in Russia, in our cause, in our strength, in our leadership.'[29]

Another image common among Gallipoli veterans was that their stay on the peninsula was a 'liturgy of the faithful', a term that draws upon Orthodox Christian theology and its division of the liturgy into three phases, with the administration of the sacraments confined to the third, most sacred, liturgy of the faithful.[30]

But the strongest religious image that was associated with the Gallipoli internment was that of the Passion and Resurrection of Christ. 'Last year', wrote the novelist Ivan Lukash in November 1921, 'everything was finished on this day. Last year, the coffin containing Russia was nailed shut and thrown into the sea. Everything was finished . . . ' Perhaps as a layman, Lukash did not attempt to take the analogy too far, but for the chaplain of I Army Corps, Reverend Fedor Milianovskii, the image of Gallipoli was clear:

> You, Gallipoli, are the summit of the new Russian Calvary[. . .] And, just as the Calvary of Jerusalem gave way to the world's feast of feasts and the world's holiday of holidays, the Resurrection of Christ, so too, following the Calvary of Gallipoli, where the Russian army was crucified, we will have a Russian feast of feasts and a Russian holiday of holidays: our Russian resurrection.
>
> This is what we Gallipolians live for, this is what we believe and what we hope for. In this lies our strength and fortitude.[31]

The writer Aleksandr Amfiteatrov drew the same analogy: Moscow, he wrote, was the symbol of Russia's death, while Gallipoli was the symbol of its resurrection. Journalist Sergei Krechetov used another biblical reference to describe Gallipoli: many are called, but few are chosen.[32] Throughout the Gallipoli literature, references to the Crucifixion and Resurrection abound: on the seventh anniversary of the internment an anonymous participant wrote that 'The Valley of Death [that is, the nickname of the encampment on the Gallipoli peninsula] became the Valley of the Resurrection.'[33] Wrangel himself often referred to Gallipoli as 'the path of the Cross' that transfigured those who trod it.[34]

The second symbol was the monument erected on the main cemetery. The monument, the cemetery itself and the very form that death took in the camp were a new phenomenon. As one participant recalled it,

> Death in Gallipoli once again became a great mystery. The deceased no longer lay with outstretched arms, stuck in the earth like some unnecessary lump, around which warfare raged like a windstorm. The deceased now lay in a coffin, surrounded by candles, while the choir sang 'Mourning over the Coffin.' And this was strange. We had gotten unused to the grandeur [blagolepie] of death. [. . .] There was a growing feeling of sorrowful respect for the dead. We didn't want to see nameless graves, which tomorrow would be ploughed over with the earth.[35]

Great care was taken to mark the graves and to ensure that a lasting monument would be created to honour the memory of the dead.

The artistic idea for the structure was proposed by sub-lieutenant N. N. Akat'ev (a former student of the Imperial Academy of Arts) and consisted of a stone mound in the Syrian style. General Kutepov issued an order that each officer and soldier, no matter what their rank or position, should contribute one stone to the monument, as was the custom in ancient times. The result was that between 9 May and 16 July 1921, 24,000 stones were collected (5000 supplementary ones in addition to the 19,000 provided in the prescribed manner).[36] The cemetery had been created on ground where it was presumed that Russian soldiers had been buried before: prisoners of the Crimean War and before them Zaporozhian Cossacks who had fought the Turks and raided Ottoman holdings in the sixteenth and seventeenth centuries. The inscription on the monument reflected this legend, which, in the absence of concrete historical evidence, must remain yet another myth that served the purpose of reinforcing the faith of the Gallipoli veterans that they were not alone, but rather formed a link in a historic chain.[37]

The monument and cemetery were cared for by the Russians until the final evacuation in 1923, and thereafter by the Turkish municipality, but an earthquake in 1939 destroyed the monument. It was only after the Second World War that the veterans became aware of this fact, however, and they resolved to build a new monument – a smaller version of the Gallipoli monument – on the Gallipoli section of the Russian cemetery at Sainte-Geneviève-des-Bois, outside Paris.[38]

At the opening ceremony of the new Gallipoli monument, on 2 July 1961, the chairman of the chief directorate of the Gallipoli society, Colonel Gorbach, sadly noted the changes of the past 40 years: 'Gathered around in uneven rows are the gray last Mohicans of Old Russia.'[39] As death took more and more of the remaining Gallipoli veterans, especially after the Second World War, their attitude towards the passing showed clearly. Several cemeteries around the world have Gallipoli sections, and the veterans' journals and bulletins always reported the deaths of one of their number: another attempt to ensure that the memory of their comrades would be retained. Indeed, aside from the stone monument at the Sainte-Geneviève-des-Bois cemetery, the Gallipoli veterans left numerous literary monuments in their wake in the form of periodicals, books and anniversary publications.

In order to maintain the spirit that had been developed during the internment (and also to secure cohesion for members of the army in case they should be called to the flag once again), already on the first anniversary a veteran's group, the Gallipoli Society, was formed. After the resettlement of the Russian army, primarily in the Balkans and

Czechoslovakia, the Russian high command in Serbia took measures to ensure that the Gallipoli spirit was maintained in every corner where veterans might be found. Sections of the Gallipoli Society were formed in every city and town where more than a few veterans settled: Belgrade, Sofia, Budapest, Prague, Brno, Brussels, Paris, Grenoble, New York, Los Angeles; even Paraguay had a society.[40]

A special role was given to the Gallipoli Society among other organizations of veterans of the White armies:

> Ideally, the Gallipoli Society must serve as the nucleus of the active element of the Russian All-Military Union [an umbrella organization encompassing regimental, naval and other societies formed by members of the White armies abroad], as opposed to those units which are bound by their charters and their mission to preserve the cadres of the Russian Army, their spirit and traditions. The active element of these units must be soaked up by the Gallipoli Society, so that it shall serve as the source of strength for all forms of struggle against the Bolsheviks.[41]

In many respects, the Gallipoli Society fulfilled the image that Anton Kartashev had received following his visit to Gallipoli: that this was an order of knights.[42]

When one examines the names of those involved in active operations against the Bolsheviks on Soviet territory in the 1920s and 1930s, and later those volunteering for the Spanish Civil War and Russian Corps (formed among émigrés in Yugoslavia by the Germans during the Second World War), it is striking that many – but not all – were Gallipoli veterans. Such was Captain Viktor Larionov, responsible for the 1927 bombing of the Leningrad Communist Party Club.

'Our campaign is not finished,' affirmed the editorial of the newspaper *Gallipoliets* in 1927, 'and the day of the final decisive battle is drawing ever nearer.'[43] Going into the Second World War, with its illusory possibility for the liberation of Russia, *Gallipoliiskii viestnik*, the organ of the largest regional society (in Bulgaria), appealed to its members to close ranks and renew their contacts with the organization: 'We CANNOT lose each other from sight', it insisted.[44] After the havoc of the Second World War (with its destruction of the Bulgarian and Yugoslav émigré centres), the military emigration emerged numerically, organizationally and morally weakened; many veterans societies and unions disappeared, but the Gallipoli Society remained intact, continuing to reject defeat, but conducting its struggle with the pen – on the pages of magazines – rather than by arms.

The myth of Gallipoli, then, was that it served as sacred locale where a morally and spiritually superior warrior was created, one called upon by example to unite and lead the Russian emigration. In fact, however, the picture was more complex. The annual report of the Regional Gallipoli Society in Bulgaria for 1928 revealed that of 600 Gallipoli veterans living in Sofia, only 171 had joined the Society, and of those even fewer were paying their dues and fulfilling other obligations. 'If not for the consciousness of the necessity of the continuing existence of the Society . . . with such the attitude of it members, there is nothing left but to close it', pessimistically concludes the report.[45] In fairness, it must be added that this was the beginning of a worldwide depression, and the figures for the previous year were not as bleak: the entire Gallipoli Society numbered over 12,000 members in 1927.[46] Aside from its goals of uniting the Gallipoli veterans in spirit and issuing various publications celebrating the ordeal, the society performed a mutual aid function, providing aid to the ill and helping locate work for the unemployed. This was particularly important during the period of the Great Depression, when regional sections of the society acted as agents in finding work for their members, sometimes even placing large groups onto works projects or, failing that, organizing their transfer to other countries where work was more readily available.[47] This activity, in turn, further consolidated the Gallipoli veterans around their society.

As is usual, the passage of time smoothed or erased the more bitter of memories, and Gallipoli became ever more of a symbol, whose hungry days, cold nights, tattered clothes, disease and filth receded into forgetfulness:

> How many wonderful nights we spent in the tents singing dear lovely songs, reading hard-to-obtain books under the meager light of flickering candles, arguing heatedly, exchanging opinions, and burning with a limitless love for the motherland we had left. [. . .] Material needs we had none (due to the enforced fast), our souls were cleansed through deep sufferings, and they began to feel more acutely and deeply. Hungry, deprived of the joys of earthly life, we somehow felt more sharply the vanity of earthly wants. We believed that the Lord was cleansing us for some great deed and that we are still needed for some thing, not knowing what it is. We believed that our motherland could not have perished for all time, that the Lord was allowing it to suffer and cleansing it by fire and harsh tests in order to once again raise it on high, where it would shine for the world to see.[48]

The teleological implications of the Gallipoli experience appeared equally clear 40 years after the events, and even as they entered old age, the veterans of the Gallipoli experience still valued that time as imbued with a spiritual meaning which had yet to be fully revealed. At the same time, there was a very real intensity added to their memory as the veterans saw world events justifying their early anti-communism.

As the émigrés aged and entered the 1970s and 1980s, the Gallipoli Society became ever more vulnerable to the physical infirmity and death of nearly all its members. Nevertheless up to the very end, the Gallipoli spirit infused these men and gave them an outlet for their hopes and activity into advanced old age. 'The life of Gallipoli veterans in Los Angeles is quiet, but difficult,' wrote the information bulletin *Gallipoliets* in 1959, 'most of them arrived in America already advanced in years, which forces them to take poorly paid work, just to scrape by. A lack of knowledge of the language makes us even less valuable. But the main thing is that the spirit of Gallipoli bonds us and holds us together.'[49] By this time, of course, the Gallipoli veterans had all exited the age of physical continuation of military struggle. Yet new reasons for the safeguarding of the Gallipoli spirit came to the fore: the regional societies helped their members retain a sense of identity and consequently kept them from slipping into 'unsightly Philistinism' and moral turpitude. Moreover, as the physical distances between the Gallipoli veterans grew (post-World War II émigrés from Europe settled in North and South America, Morocco, Australia and other places, and were separated by large distances even within cities as they settled in suburbs), the Gallipoli Society tended to become an extended family, offering its members moral and material support in addition to the fondest of memories.

As one historian has recently noted, 'Whether or not Gallipoli really was the special experience that was claimed, many certainly came to believe that it was. In a sense the mythology made it real.'[50] The efforts of Wrangel and Kutepov transformed the defeat of the Crimea into the victory at Gallipoli for a large number of people for the length of their lives. It has been argued that in so doing, the leadership of the White Army made it more difficult for them to assimilate and become engaged citizens of their new host countries, but this begs the question of how assimilable former military men, most of whom were reduced to physical labour, could be. The myth of Gallipoli could equally well be seen as a rescue line, giving men who might otherwise sink into depression, alcoholism, or commit suicide, a reason to live and to hope.[51] For people who had lost their homes, their possessions, their families and their motherland, Gallipoli was the only memory that could not be taken away or besmirched. But it was also a memory that was difficult to transfer to future generations or non-participants. There was no nation to receive this memory, process it and reuse it, so that the Gallipoli myth, powerful as it was, could not outlive its veterans.

Notes

1. N. Karpov, *Krym – Gallipoli – Balkany* (Moscow: Russkii put', 2002), 11.
2. Petr N. Wrangel, *Vospominaniia* (Frankfurt: Possev-Verlag, 1969) vol. II, 237.
3. V. Kh. Davatts and N. N. L'vov, *Russkaia armiia na chuzhbinie* (Belgrade: Russkoe izdatel'stvo, 1923), 14.
4. Reports dated 18 and 19 December 1920, Hoover Institution Archives (HIA), Wrangel collection, 145.28.
5. *Russkie v Gallipoli, 1920–1921* (Berlin, 1923), 41. Others set up their new homes in caves at the shoreline, mausoleums at the cemetery, in dried-out water reservoirs, and anywhere else there was a semblance of shelter.
6. See the Wrangel collection, 89.15, for materials relating to the application of the death penalty.
7. Undated memorandum regarding items that should be explained to the troops, HIA, Wrangel collection, 145.28.
8. Order of the commander-in-chief of the Russian Army dated 5 January 1921, HIA, Wrangel collection, 152.48.
9. V. K. Vitkovskii, *V bor'be za Rossiiu: Vospominaniia* ([San Francisco], 1963), 30–6. Vitkovskii himself felt that the plan was pure adventurism, but, he added, was not Hannibal's crossing of the Alps also a gamble?
10. V. Matasov, 'Konnoartilleristy v Gallipoli' in *Russkaia armiia na chuzhbine: Gallipoliiskaia epopeia* (Moscow: Tsentrpoligraf, 2003), 228–9.
11. G. Orlov, 'Iz dnevnika gallipoliitsa', *Pereklichka*, 109 (December 1960), 3. This was not the first instance of conflict with the French garrison: in the first month of the Gallipoli internment, a French patrol arrested two officers, and when the French commandant refused to hand them over to the Russians, two companies of armed junkers (military college students) were sent to free them; this they accomplished without shooting (the Senegalese abandoned their positions and two machine guns), and no more Senegalese patrols were sent into town.
12. B. Esipov, 'General Kutepov', *Soglasie*, 101 (February 1960), 19.
13. Circular from Wrangel, 10 May 1921, HIA, Wrangel collection, 145.28.
14. General Wrangel issued an order to this effect on 26 September 1921.
15. Letter of Lieutenant General P. A. Kussonskii (temporary chief of staff to General Wrangel) to the Russian military attaché in Greece, 8 September 1921, in 'Institut voennoi istorii Ministerstva oborony Rossiiskoi Federatsii', *Russkaia voennaia emigratsiia 20-kh – 40-kh godov: Dokumenty i materialy*, vol. I, pt II (Moscow: Izdatel'stvo 'Geia', 1998), 62.
16. Boris Mirskii, 'Gallipol'skaia anketa', *Dni*, 158 (9 May 1923). Mirskii never visited Gallipoli. On Gallipoli as one of the main apples of discord in émigré politics, see K. Zaitsev, 'Armiia. Neskol'ko slov po povodu knigi V. Kh. Davattsa i

N. N. L'vova "Russkaia armiia na chuzhbinie"', *Russkaia mysl'* (Prague-Berlin), VI–VIII (1923), 413–14.

17. Gallipoliets, 'Golos Gallipoliitsa', *Obshchee dielo*, 323 (4 June 1921). This article became an object of animated discussion within the camp. See G. Orlov, 'Iz dnevnika gallipoliitsa', *Pereklichka*, 109 (December 1960), 5.
18. Ivan Lukash, 'Liturgiia viernykh', *Rul'* (Berlin) 304 (16 November 1921).
19. G. Orlov, 'Iz dnevnika gallipoliitsa', *Pereklichka*, 109 (December 1960), 3.
20. Gallipoliets [pseud.] 'Sila i smysl Gallipoli', *Gallipoli, Lemnos-Chataldzha-Bizerta: Iubileinyi Al'manakh-pamiatka izdannyi k 35-letiiu prebyvaniia v Gallipoli Russkoi Armii, 1920–1955* (Hollywood, CA [1955]), 9.
21. In this regard, the situation on Lemnos may have been somewhat worse: winds and torrential rains swept the island over the winter, tearing down tents. As a result, witnesses reported a greater prevalence of illness, including serious diseases such as tuberculosis.
22. Report on conditions on Lemnos and Gallipoli by P. P. Perfil'ev, undated (approximately July 1921), in Institut voennoi istorii Ministerstva oborony Rossiiskoi Federatsii, *Russkaia voennaia emigratsiia 20-kh – 40-kh godov: Dokumenty i materialy*, vol. I, pt II (Moscow: Izdatel'stvo 'Geia', 1998), 62. Also indicative of the difference in spirit was General M. Georgievich's letter to General P. A. Kussonskii, dated 2 October 1921. Having heard that Gallipoli, Lemnos, and Çatalca were all to be treated equally in terms of a medal to be issued, Georgievich was upset: conditions were the same, he wrote, but the behaviour of the troops was completely different, HIA, Wrangel collection, 145.28. Altogether, over 8500 men left Lemnos for Soviet Russia, many times more than did so from Gallipoli.
23. S. Rytchenkov, *259 dnei lemnosskago sidieniia* (Paris, 1933), 3.
24. *Russkie v Gallipoli*, 440–1, *Russkaia armiia v izgnanii, 1920–1923* (n.p., n.d.), 8. Units of some 1500 Russians under the command of General Martynov remained in Gallipoli until 1923, when it too was evacuated to Serbia.
25. Quoted in Vasilii Goncharov, 'Goloe pole', *Rodina* (Moscow), 5–6 (1998), 136.
26. General F. F. Abramov's comments in *Zhivym i gordym* (Belgrade: Izdanie Obshchestva Gallipoliitsev, 1923), 9.
27. Vladimir Dushkin, *Zabytye* (Paris: YMCA-Press, 1983), 103.
28. 'Itogi chetyrekhletnei bor'by', *Obshchee dielo*, 499 (29 November 1921), 2.
29. Anonymous letter (signed 'R. P.') to Wrangel, dated 1 July 1921, HIA, Wrangel collection, 145.28.
30. Ivan Lukash, 'Liturgiia viernykh', *Rul'* (Berlin) 304 (16 November 1921). B. Esipov, op. cit., 20, repeats Lukash's title in his reminiscences on the fortieth anniversary. The term had also been used as the heading of the second part of V. Kh. Davatts's *Na Moskvu* (Paris: Tip. Akts. O-va I. Rarakhovskii, 1921).
31. Rev. F. Milianovskii, 'Khristos Voskrese!' *Gallipoli* (Belgrade), 2 (8 April 1923). In contrast, G. Orlov, a Gallipoli veteran, wrote that no one he knew felt themselves 'crucified' upon evacuation, G. Orlov, 'Iz dnevnika Gallipoliitsa', *Pereklichka*, 108 (November 1960), 10.
32. 'Gallipoli (anketa)', ibid.
33. Ia. R. '1920–1927', *Viestnik Gallipoliitsev v Bolgarii* (Sofia), 5–7 (September–November 1927), 2. The anniversary almanac, *Russkie v. Gallipoli, 1920–1921* (Berlin, 1923) contains innumerable examples of such associations.

34. See, for example, his order #1, 1 January 1923: 'Another year has gone by as we trod the path of the cross', HIA, Wrangel collection, 117.3.
35. Gallipoliitsy [pseud.] 'Gallipoli', *Pereklichka*, 93 (July 1959), 6.
36. 'Russkaia sviatynia v Gallipoli', *Gallipoli* (Belgrade), 1 (15 February 1923).
37. The inscription read, 'Lay to rest, O Lord, the souls of the dead. I Army Corps to its warrior brothers, who for the honor of the motherland found their final rest on foreign soil in 1920–1921, in 1854–1855, and in memory of their ancestors, the Zaporozhian Cossacks, who perished in Turkish captivity.'
38. For a while in the 1920s, Ismail Ikhsan Oglu (Russian spelling) served as cemetery guardian, reporting to Russian officials in Constantinople, but in the 1930s his reports ceased, and so too, presumably, did his service. Altogether, 274 Russians were buried there. The story of the fate of the monument is told by V. V. Polianskii, 'Gallipoliiskii pamiatnik', *Obshchestvo Gallipoliitsev* (Paris: Izdanie Glavnago Pravleniia Obshchestva Gallipoliitsev, 1953), 5–7.
39. *Gallipoliets: Informatsionnyi biulleten' Glavnago Pravleniia Obshchestva Gallipoliitsev*, 83 (22 November 1970), 4.
40. Interestingly, even the local Gallipoli societies as such were capable of creating a Gallipoli-like atmosphere and became themselves objects of memory. See, for example, Vladimir V. Al'mendinger, *Gallipoliiskoe zemliachestvo v Brno: Pamiatnaia zapiska o zhizni gallipoliitsev v Brno (Chekhoslovakia), 1923–1945* (Huntington Park, CA: Author, 1968). A 'Gallipoli church' was founded in Paris, which contained many of the artefacts from the field churches on Gallipoli. It appears to have functioned into the 1950s, at least *Informatsionnyi biulleten' Otdela O-va Gallipoliitsev v S.A.S.Sh-kh.*, 48a (15 July 1955), 3–4.
41. Excerpt from a letter of General I. G. Barbovich, published in the proceedings of a meeting of the members of the Gallipoli Society in Czechoslovakia, 4–5 June 1933, 31, HIA, Chasovoi collection, Box 1. When the units that had been transferred to Bulgaria began to disperse, as individuals sought better opportunities in France, Belgium, and other Western European countries, General Kutepov issued a directive calling on commanders of units of I Army Corps to appoint elders in regions where their officers and soldiers could be found to maintain contact with them in the interests of 'securing our traditional Gallipoli bond', directive dated 14 September 1923, HIA, Wrangel collection, 145.29.
42. 'Doklad prof. A. V. Kartasheva "Vpechatleniia o poezdke v Gallipoli", prochitannyi na sobranii, ustroennom Komitetom Londonskogo Otdiela Natsional'nago Soiuza 25/XII 1921 g.', HIA, Nicolaevsky collection, 16.1.
43. *Gallipoliets* (Paris), 1 (1927).
44. *Gallipoliiskii viestnik* (Sofia), 84 (1 June 1940), 16.
45. 'Godovoi otchet Oblastnogo Pravleniia Obshchestva Gallipoliitsev v Bolgarii za 1928 god', 10 April 1929, 4, HIA, Mariia Wrangel papers, 11.3. Nearly two years later, the leadership of the Regional Directorate was repeating this refrain, op. cit.
46. M. I. Rep'ev, 'Gallipoliets i ego obshchestvo', *Gallipoliets* (Paris), 1 (1927).
47. See, for example, 'Godovoi otchet Oblastnogo Pravleniia Obshchestva Gallipoliitsev v Bolgarii za 1928 god', 10 April 1929, 4, HIA, Mariia Wrangel papers, 11.3.
48. Z. Slesarevskaia, 'Sviataia noch' v Gallipoli', *Pereklichka*, 101, (April 1960), 3.

49. *Gallipoliets* (Paris), 32 (February 1959), 3.
50. Paul Robinson, *The White Russian Army in Exile, 1920–1941* (Oxford: Clarendon Press, 2002), 49.
51. Just one example of how hard the lot of these former military officers was: a report from Nice, France, early in 1941, states, 'last week in Cannes de la Bocca, 56 men from Col. Tarasevich's group, working at the garbage dump, were laid off work. Most of them were Gallipoli veterans. I have no idea how to help.' One can only wonder what level of ideals must be necessary to fill the souls of men who should have been serving their country in their chosen profession, yet instead were working at a garbage dump in a foreign land. The fate of many others – miners in Belgium and Bulgaria, construction workers in Yugoslavia, factory labourers in France – was not much better.

13
The Memory of French Military Defeat at Dien Bien Phu and the Defence of French Algeria

Stephen Tyre

> To lose an empire is to lose yourself. It takes all the meaning away from the life of a man, the life of a pioneer.[1]

On 7 May 1954, the French garrison at Dien Bien Phu, in northern Indochina close to the Laotian border, fell to the forces of the nationalist *Viet Minh* under General Giap, after two months' fighting. The Vietnamese victory – widely seen as the first military defeat of a European colonial power by an anticolonial guerrilla force – hastened the French withdrawal from Indochina, which had already been anticipated by the opening of the Geneva Conference on the future of the colony, on 26 April. The Geneva Accords, which formalized the ceasefire on 21 July 1954, did not therefore come about solely because of the French defeat at Dien Bien Phu. Nonetheless, this fact did not lessen its impact both in France and worldwide. In France, the defeat was a significant factor in the fall of yet another government of the unstable Fourth Republic, when Prime Minister Joseph Laniel was succeeded, on 19 June, by Pierre Mendès-France, elected by the National Assembly with an unusually large majority, thanks to his promise to negotiate an armistice in Indochina. The nationalist Right and elements of the officer corps, who had grown increasingly disillusioned with the fragile coalitions and consensus politics of the Republic, seized upon both the defeat at Dien Bien Phu and the subsequent armistice to accuse the 'regime' or the 'system' – as the Right were wont to describe the Fourth Republic – of abandoning vital French interests. Internationally, two of the most significant repercussions of Dien Bien Phu were the deepening of the American commitment to Indochina and the encouragement that the French defeat offered nationalists in France's North African territories of Morocco, Tunisia and Algeria.

When the Algerian War of Independence began on 1 November 1954, the Algerian nationalists' understanding of Dien Bien Phu's importance was matched by that of the French Right and the military.

In spite of Dien Bien Phu's clear importance as a turning point in French decolonization, recent historical works on the dissolution of the French empire and on the memory and legacy of French colonialism and decolonization have focused on Algeria to the detriment of Indochina. Just as the outbreak of war in Algeria less than six months after the defeat in Indochina restricted the potential for considered reflection on the lessons of Dien Bien Phu, so the much greater French military commitment to and loss of life in Algeria has tended to obscure Dien Bien Phu's unique and decisive contribution to the mental world of the veterans of France's wars of decolonization. Memories of the Algerian War have been extensively studied and memories of Indochina relatively overlooked in comparison,[2] but as this chapter aims to demonstrate, many veterans' and politicians' memories and judgements about the significance of the French war in Algeria were shaped in important ways by the way in which their memories of the defeat at Dien Bien Phu had shaped their attitudes to the subsequent conflict in North Africa. In the post-colonial period, memories of Dien Bien Phu evolved, in the light of the defeat in Algeria, and changed in nature, moving from personal memories of defeat and humiliation to collective ones of national decline in which Dien Bien Phu was often seen as a starting point.

Considerable attention has been paid to the difficulties which both the Second World War and the Algerian War posed for the evolution of a national collective memory in France. In the case of the Second World War, historians have identified a collective will to forget the less palatable aspects of the war as experienced by the French, resulting in the development of a so-called resistance myth which began to lose credibility in the 1970s. Since then, the colonial war in Algeria might be said to have taken up Vichy's mantle as the latest of France's best-forgotten twentieth-century wars. For many historians, this undeclared 'war without a name' continued, until the end of the 1990s, to be marked not by a collective memory but rather a collective amnesia. Since around 1999, however, veterans and victims of the Algerian War have succeeded in bringing the memory of the war into national focus, resulting in a proliferation of memoirs and historical studies, accompanied by some high-profile legal cases.

The war in Indochina, unlike the Second World War and the Algerian War, was a campaign fought in a distant theatre, by an expeditionary force composed entirely of career soldiers; in spite of the French command's conviction that French forces were numerically inadequate against the manpower resources of the Viet Minh, no French government was prepared to commit conscripts to Indochina. In Algeria, on the contrary, reservists and conscripts participated in operations from 1956 until 1962. Indochina was, therefore, an experience shared only by professional soldiers. The defeat at Dien Bien Phu was not entirely devoid of resonance in the wider French population; news reached France at the time of the VE day commemorations in May 1954. However, these commemorations were themselves the object of some controversy in post-war France, with 8 May only having been officially designated a national holiday and day of commemoration in France in 1953.[3] Furthermore, within months of Dien Bien Phu, the war in Algeria occupied French attention, thus filling the space that might have been reserved for contemplation of the Indochinese defeat. Dien Bien Phu did not become invisible in the immediate aftermath of the defeat, but a certain version of events in Indochina, filtered through the war in Algeria, quickly rose to prominence, as the so-called colonial party, comprising professional soldiers, the political Right, and settler populations in the colonies, emerged as the group with most interest in keeping its memory of this colonial defeat alive, with the intention of preventing any repeat of the events of May 1954.

This study will therefore take as its focus these interest groups on the pro-colonial French Right and among the military. It aims to show how a shared memory of Dien Bien Phu was shaped in the immediate aftermath of military defeat, and reinforced in turn by the emergence of consensus among these elements about the lessons to be drawn from that defeat. The consensus among these groups regarding their memories of Dien Bien Phu, which developed against a background of indifference or ignorance among the majority of the French population, therefore allowed many individuals' conclusions about Dien Bien Phu's importance to shift from having personal to national significance. The trauma of May 1954 grew, in many cases, from a personal defeat to become, by the 1970s, the beginning of a national humiliation, as the veterans' memory was shaped by the war in Algeria and the American War in Vietnam. This study of the evolutions of veterans' memories of their defeat in Indochina therefore provides an example of how a collective memory of a defeat and its causes, when nurtured and reinforced by a closely defined group sharing a series of formative experiences led individuals to adopt extreme positions and to amplify the significance of their memories of Dien Bien Phu.

The emergence of a rather politicized memory of Dien Bien Phu among sections of the French Right owes much to the lack of more widespread consensus on the reasons for and the scale of the defeat, and on its consequences. For some observers, it was reasonable to argue that Dien Bien Phu was not really a disaster for France. For Pierre Messmer, a serving officer in 1954 and later Minister of Defence and Prime Minister, many memories of the battle seemed to have overlooked its less catastrophic aftermath:

> The heavy defeat at Dien Bien Phu didn't lead to a debacle. Only rarely do people speak of the 'military revival' that followed, which allowed the French troops, their Vietnamese allies and the Catholic population to leave North Vietnam freely and under normal conditions during the following year.[4]

However, in the absence of any national commemoration or consensus in the years following Dien Bien Phu, conflicting interpretations about the defeat and its importance were able to flourish. For another soldier who went on to become a Gaullist politician, Alain Griotteray, events away from Indochina made the military defeat more serious than it needed to be:

> Even after the fall of Dien Bien Phu, it wasn't all over. The army had written another epic page in France's amazing military history, an adventure as moving as Napoleon's last battle in France in 1814. Dien Bien Phu had been nothing but an unfortunate failure. The expeditionary corps had seen others like it. The general debacle which some were predicting after the defeat hadn't happened. The defeat hadn't diminished the resolve of the French troops; on the contrary, they were determined to make amends for it.[5]

In this case, Griotteray suggests, the emerging memory of disaster must have been shaped by something; the most likely explanation, as will be seen below, is the political context in which the battle took place. It is significant, too, that both these interpretations assert that there was no 'debacle' in Indochina; the word evokes memories of the Fall of France in 1940 and reminds us that one reason for the resonance of Dien Bien Phu on the nationalist Right is that the defeat could be seen as adding a new layer to an existing memory of humiliating defeat tinged by political incompetence, which had taken hold in France since 1945 in relation to the 1940 defeat.[6]

Messmer and Griotteray's insistence on the wider context of Dien Bien Phu is typical of the views expressed by many veterans and their sympathizers on the political Right. Veteran Paul Boury asserts that

'Although just a battle, but a cruel and heroic battle, Dien Bien Phu is nevertheless the start of a whole new period in French history',[7] while another veteran, Roger Bruge, describes it as 'the last battle fought by the French army'.[8] Jean-Yves Alquier, an officer best known for his commitment to the French 'pacification' campaign in Algeria, recalls Dien Bien Phu in terms of political betrayal in Paris as well as military defeat:

> A few men are fighting for freedom against a barbarian horde and, in the capital cities of the Western world, which have at their disposal the military means to make the enemy retreat, discussions are still going on about whether to help them or not. That's a terrible situation. It's one of the darkest pages in the history of the free world.[9]

Memories of Dien Bien Phu as an avoidable disaster which revealed the failings of the ruling political class in France have been kept alive by others, who were not present in Indochina but followed events from Paris with sympathy for the fate of the defeated French troops. Historian Raoul Girardet, who was active in Right-wing nationalist politics during the Algerian War, recalls Dien Bien Phu as an important turning point in his own political commitment:

> Dien Bien Phu was a decisive shock. Once again, humiliation, defeat, lines of prisoners, the flag being lowered, the abandonment of loyal populations . . . It was on the Sunday after the announcement of Dien Bien Phu's capitulation that I got into a fight in the street for the first time.[10]

Likewise, *Front National* politician and Dien Bien Phu veteran Jean-Marie le Pen echoes Girardet's memory of Dien Bien Phu as a decisive moment which determined subsequent political action, recalling that

> We arrived too late to jump into the centre of hell with our mates. We all cried tears of grief and rage, on that evening of May 8. It was on that day that I understood that to win wars you need more than soldiers and courage, that wars are won and lost away from the battlefield. I swore that if I made it back, I would devote my life to politics.[11]

This range of memories of Dien Bien Phu as something more than a simple military defeat, frequently interpreted as a call to action, goes a long way to explaining how this defeat, despite the relative lack of attention paid to it by the French population at the time and by historians of memories of war and defeat in twentieth-century France, has nonetheless been remembered, and interpreted as a decisive and politically significant event, by an important body of opinion on the French Right. In what follows, the immediate context in which memories of Dien Bien Phu were shaped will be examined, with the aim of establishing the all-important 'social frameworks' in which these memories evolved. Subsequent sections will look at the transmission and evolution of memories of Dien Bien Phu through the Algerian War and the American War in Vietnam.

Immediate reactions to defeat

Many studies of memories of war in the twentieth century emphasize the role of 'social frameworks' in shaping the evolution of memories shared among groups with similar experiences and outlooks.[12] For the veterans and politicians discussed above, the immediate political responses to the defeat can be seen to have made an important contribution to the crystallization of an emerging memory of a defeat which was due to political neglect as much as military errors, and which ought to be seen as a call to action if France was not to suffer the same fate in the rest of its colonies. As early as 1955, an official government Commission charged with explaining the defeat concluded that

> The Commander in Chief didn't feel supported. He had reason to believe that the war in Indochina was being inadequately directed and followed in Paris. He knew that France had almost run out of resources, and that the parsimony with which he was granted reinforcements and the slow pace at which they reached him, caused mounting difficulties in carrying out his instructions.[13]

These immediate findings played a key role in shaping the collective memory and 'lessons' of the defeat, suggesting which memories and interpretations of Dien Bien Phu would pass into veterans' and the Right's collective memory. The press, too, was influential in depicting Dien Bien Phu as a case of a brave fighting force being defeated thanks to a lack of political support. After an initial depiction of heroic defeat – *France-Soir* announced that 'the epic of Dien Bien Phu proves to the world that heroism is still one of the profound virtues of the French'[14] – the media quickly embarked upon a search for guilty parties, fuelled by a conviction that the heroes of Dien Bien Phu had been let down by politicians in Paris and Washington. To this was quickly added a sense of moral weakness on the part of the French themselves; *Le Figaro* claimed that 'Those who fought at Dien Bien Phu died because we lied to ourselves . . . Opportunities for victory were lost . . . out of weakness.'[15]

By 1956, the tendency was to see Dien Bien Phu in a wider perspective, placing the emphasis on its value as a symbol of pointless sacrifice and impending catastrophe. On the second anniversary of the defeat, in the first example of special media coverage devoted to an anniversary of the battle, the popular and influential *Paris-Match* delivered the following analysis:

> The defeat at Dien Bien Phu wasn't just a military episode, a redeemable failure whose useless dead are quickly forgotten. It's the key date on which the process of disintegration of the French empire began. Until then, France still had a reserve of prestige. Never had she been beaten by the peoples whom she protected . . . And there France lost the test of strength which she had accepted to undergo. There the French officers, heroes of so many campaigns, heirs to so much culture and so many victories, were taught a lesson by those little, lazy yellow men whom they considered, in 1940, barely capable of fighting alongside them.[16]

The analysis offered here contains most of the themes that emerged prominently in veterans' and the Right's memories of what had happened at Dien Bien Phu: this had been more than a simple military defeat; the French deaths had been in vain; it should be seen as the beginning of the end of France's colonial 'mission', even though by May 1956 only the protectorates of Morocco and Tunisia had followed Indochina in gaining independence; and it was to be compared and contrasted with the way in which France had been able to overcome the humiliation of defeat in 1940. The implication that at Dien Bien Phu the French army came up against a new kind of enemy and a new force which it seemed unable to defeat further reinforced the notion of this defeat as a decisive turning point in the loss of French imperial prestige, which dominated recollections of the battle's importance in the 1950s.

Dien Bien Phu also became a political weapon in the remaining four years of the Fourth Republic. Immediate reactions to the defeat on the Right predictably called for the government's resignation, such as the editorial in the Catholic newspaper *l'Aurore*: 'The lesson of their sacrifice is *National Union* in the face of the Communist war.'[17] The chief opponents of the Fourth Republic on the Right were the Gaullists, who refused to participate in the coalition governments of the republican 'system' while campaigning for the return to power of General de Gaulle and the implementation of a new constitution with a strong executive. Throughout the period from May 1954 to May 1958, the memory of Dien Bien Phu was frequently used by Gaullists to signify a humiliating defeat in the face of world opinion or a moment of crisis for the government in France.[18] Already, on 7 April, with the French garrison at Dien Bien Phu under siege and many predicting a humiliating defeat, de Gaulle had staged a high-profile press conference in which he called for a mass show of support for his quest to restore French prestige, on the occasion of the VE day commemoration at the Arc de Triomphe in Paris:

> I ask the people to be there to show that they remember what was done to save France's independence and that they intend to keep it. I ask veterans of the two World Wars and the Indochina War to surround the monument . . . All of us, however many we are, who are present on that day, will remain silent. Above this solemn silence will hover the soul of the nation.[19]

By the time of the ceremony on 9 May, the fall of Dien Bien Phu had added extra urgency and significance to de Gaulle's somewhat melodramatic appeal. The link between humiliation in the colonies and a decisive rallying around de Gaulle now clearly being made by Gaullist activists, who even planned to turn the commemoration into a *coup d'état* for which an association of Indochina veterans, disillusioned by Dien Bien Phu, would provide the muscle and a combination of Gaullists and the Second World War commander Marshal Alphonse Juin the political leadership. The coup was abandoned, to the annoyance of some of the veterans' groups, when the turnout at the Arc de Triomphe turned out to be less than had been hoped. However, this demonstration of May 1954 has been interpreted as a forerunner of the similar events of May 1958 which did bring about the return of de Gaulle and the end of the Fourth Republic. It demonstrated that a mass rally around de Gaulle could capture the imagination of activists and military figures, but also that the defeat in Indochina did not constitute a sufficiently serious crisis, in the eyes of the general population, to warrant direct action to overthrow the regime.[20] Thus, the potentially powerful but rather limited political application of the veterans' and the Right's memory of

Dien Bien Phu, at this very early stage, was revealed, leaving the Gaullist activist Jacques Soustelle to conclude with reference to the French defeat at the hands of Prussia in 1871 that 'Once a regime has had its Mexican War, it's perhaps not far from its Sedan, and we've come to the stage of asking how much time is left before the new Sedan which the events in the Far East seem to foretell.'[21]

Despite the Right's failure to take advantage of Dien Bien Phu to overthrow the Fourth Republic, existing links between Gaullists and veterans were reinforced by the involvement of both groups in the preparation of the demonstration of May 1954. Thus, the 'instant lessons' of Dien Bien Phu and the memory of humiliation and betrayal emerging among both veterans and politicians seemed to provide a potential justification for radical political action in the name of the recent military defeat.

A further important thread running through political reactions to Dien Bien Phu was the emergence of a 'stab in the back' theory, which also found its way into the emerging collective memory of the defeat. The guilty party in this respect could be the Communists, as *Le Figaro* asserted, 'The real winners tonight are the friends of [French Communist Party leaders] Mr Thorez and Mr Duclos. It's they who should be raising the red flag of death over the ruins and the tombs.'[22] Alternatively, blame for Dien Bien Phu could be attributed to the United States, whose lack of military support for the French was frequently cited by frustrated French commanders as an explanation for the deteriorating military situation in Indochina.[23] The recollection of veteran Jean-Marie Juteau is typical of veterans' memories of lack of American support at Dien Bien Phu:

> They treated us with contempt, as if we didn't even exist. Some of them didn't even know where France was! At the time of Operation Castor,[24] the Americans could have joined with France in facing up to the Vietnamese communists. It was perhaps the only way to solve the problem, if only temporarily. But they preferred to watch us leave and do it alone.[25]

Thus, the immediate 'social frameworks' in which the emerging memories of Dien Bien Phu were processed among veterans returning from Indochina and nationalist politicians in France focused on three key aspects: the lack of political and national will to devote the necessary resources to the defence of the French empire, the potential for a coming-together of disaffected veterans and the Right to act decisively against the Fourth Republic, and the perception that France had been betrayed in Indochina by a Communist fifth column and broken American promises. All of these elements contributed to the evolution of memories of Dien Bien Phu from simple recollections of military defeat into a narrative of a decisive turning point and national humiliation. The war of decolonization in Algeria provided a second stage in the crystallization and politicization of these emerging memories of Indochina.

The Algerian War of Independence and memories of Dien Bien Phu

It is well known that Algerian nationalists sought to keep a memory of Dien Bien Phu alive in the minds of the colonized population of North Africa, to serve as an inspiration. But despite the French government's constant refusal to acknowledge that the uprising which began on 1 November 1954 in Algeria was indeed a war and the latest stage in the challenge to French colonialism, there is evidence that Dien Bien Phu veterans and their supporters quickly realized the importance of their experience in Indochina to the situation in Algeria. Many soldiers who fought at Dien Bien Phu were sent to Algeria after a brief stay in France. One such soldier recalls that his memories of Dien Bien Phu began to make sense in the light of the war in Algeria: 'I was too young, too carefree, and not mature enough to analyse the Indochina War; I understood a few years later in Algeria.'[26] Among troops drawn from the European settlers of Algeria, the relevance of Dien Bien Phu to the new conflict was more immediately apparent; not only were these veterans warmly welcomed back to their home ports, compared to the indifference and in some cases outright hostility which greeted soldiers returning to France from Indochina,[27] but reports on troops' morale reveal that these troops were already convinced, upon returning to Algeria, that what they had witnessed at Dien Bien Phu would have serious consequences for 'France's place in the world and the French Union'.[28]

In July 1955, a parliamentary delegation was sent to report on the situation in Algeria with the explicit aim of finding out how to avoid a repetition of the errors of Dien Bien Phu. The lessons of this defeat, as seen by those veterans now serving in Algeria, were of 'too much suffering for a final Communist victory'[29] and 'a war lost in Paris, the governments having rejected outright any solution that would have allowed for independence while retaining some French presence which was in any case accepted by the majority of the native population'.[30] Pressed into service by governments anxious not to commit the same errors in Algeria as in Indochina, Dien Bien Phu became an ever-present memory and a national disaster never to be allowed to recur: during the war in Algeria it was a symbol of defeat, abandonment, betrayal, humiliation. As will be seen in the rest of this section, the contribution of the memory of defeat in Indochina to the war in Algeria must be seen in these terms.

One of the first aspects of veterans' memories of Dien Bien Phu to be translated into reflections on the war in Algeria concerned the lack of political support. In 1957 the commander-in-chief in Indochina at the time of the defeat, General Henri Navarre, published his memoirs, in which he alleged that

> We had no policy at all. After seven years of war, we had ended up in a total impasse and no-one, from the rank-and-file troops to the high command, really knew why we were fighting . . . This uncertainty about our political aims is what prevented us from having a coherent military policy in Indochina.[31]

Published during the crucial Battle of Algiers, the account of the former commander in Indochina played a key role in keeping this memory of aimlessness and abandonment alive in military circles. Those who wished to see their struggle in Algeria as part of a larger problem or a longer-term movement found support for their views in the memoirs of Navarre, who also sought to shift the moment of defeat away from the battle at Dien Bien Phu towards the negotiating table at Geneva: 'It's from Geneva, not Dien Bien Phu, that France's humbling dates. It's the politicians not the soldiers who ought to be held to account.'[32]

With the collective memory of a betrayal, by an incompetent political class, of an effective fighting force reinforced in this way, the deterioration in civil–military relations in Algeria during the last years of the Fourth Republic could be made to fit into a historical narrative of decline, debacle and disorder beginning in 1940 and amplified by Dien Bien Phu. Many of the officers who had developed this memory of Indochina and Dien Bien Phu were among the most ardent believers in the theory of revolutionary warfare; inspired by Maoist tactics, this doctrine proposed a politicization of the military effort, a total immersion of the soldier among the population and the use of 're-education' to win the war of ideas.[33] The link between officers' memory of experience (often captivity) in Indochina and action in Algeria was very clear with officers frequently referring to the Algerian nationalist forces as 'les Viets' and, in many cases, developing almost an obsession with communism and a tendency to see the war in Algeria as the latest in a series of coordinated communist attacks on the Western world.[34] The politicization of those officers arriving in Algeria with their memories of Dien Bien Phu not only contributed to the breakdown in civil–military relations but was also, in the eyes of the high command, a direct and unwelcome consequence of the officers' attempt to let their memory of Indochina guide their actions in Algeria. General Cherrière complained that 'certain terms, revealed by the media, lead public opinion to make connections with certain episodes in Indochina which didn't turn out well'.[35] Thus, the memory carried from Dien Bien Phu into Algeria could actually function to the detriment of the French military effort there, encouraging a monolithic reading of colonial wars as communist-led attacks on Western civilization, leading to a loss of faith in the political leadership and perhaps also alienating public opinion from the effort in Algeria.

Yet among civilians too, memories of the political reaction to Dien Bien Phu also encouraged the advocates of revolutionary warfare in Algeria to prevent a repeat of the earlier defeat. The rhetoric of the Gaullist senator and future Prime Minister Michel Debré, for example, was heavily influenced by the reaction to Dien Bien Phu, which was itself cited as evidence that the parliamentary 'regime' was to be considered illegitimate, and unable to prevent the spread of communist rule to metropolitan France which would surely follow the loss of Algeria. In April–May 1958, France was without a government for four weeks as the Fourth Republic began to crumble under the pressure of the war in Algeria. At this stage, the Left also began to evoke memories of Dien Bien Phu for political purposes. René Pleven, who was considered a likely Prime Minister should a coalition government be formed, was constantly attacked by the Communist newspaper *L'Humanité* as 'the War Minister of Dien Bien Phu'. Meanwhile, on the day before the settler and military revolt in Algiers that finally brought down the Fourth Republic, Algerian Governor-General and Robert Lacoste were fulminating against rumours that the incoming government would open negotiations with the nationalists, insisting that there would be no 'diplomatic Dien Bien Phu' in Algeria.[36]

Looking back on the revolt that brought down the Republic in May 1958, one veteran of both the Indochinese and Algerian wars explained that

> May 13 1958 happened; that's the consequence of the battle of Dien Bien Phu. For the colonized, Dien Bien Phu has become the July 14 of decolonization. For the officers, lieutenants or captains in Algeria, it was clear that the system had to be changed.[37]

The significance of Dien Bien Phu for both sides in Algeria is clear. Yet this veteran's recollection asserts that a lost battle, in a war about which public opinion was indifferent, which was not seen or experienced by the nation as a whole, somehow led, four years later, to a coup against the Fourth Republic. The notion of Dien Bien Phu as 'the July 14 of decolonization' further emphasizes how the French colonial party saw the very future of the Republic as bound up with the legacy and memory of Dien Bien Phu. By 1961, with President de Gaulle now determined to allow Algeria to become independent, this attitude led some Dien Bien Phu veterans serving in Algeria into outright insubordination and illegality as they participated in the unsuccessful military putsch designed to overthrow de Gaulle and keep Algeria French. For Benjamin Stora, by 1961 many 'chose, in Algeria, illegality rather than a second Dien Bien Phu'.[38] One such officer was Captain Jean Pouget, a veteran of Dien Bien Phu who participated in the Algiers putsch and whose volume of memoirs, entitled *We Were at Dien Bien Phu*, was removed from circulation by the Gaullist government still nervous about military discontent following the events of Algeria.[39] For many Dien Bien Phu veterans who took part in the putsch, the chief motivation was not to repeat the 'abandonment' of 'loyal' natives that they remembered as one of the consequences of the withdrawal from Indochina:

> We remembered the evacuation of the Highlands, villagers clinging on to our lorries who fell, exhausted and crying, into the dust on the road. We remembered Dien Bien Phu, the Vietminh's entry into Hanoi . . . We remembered the villages we had abandoned and the inhabitants who had been massacred . . . We thought of all those solemn promises we had made on this African soil. We thought of all those men, all those women, all those children who had chosen France.[40]

The memory of this 'abandonment' at Dien Bien Phu not only informed certain officers' conduct in Algeria but also continued to shape the memory of the French defeat in Indochina throughout the following decades: 'Those are scenes which will forever remain in the memory of those who saw them . . . Do we need to describe what happened in the villages to which we had offered our protection, and who had accepted it, and which we abandoned?'[41]

After the end of this last war of French decolonization, Dien Bien Phu remained central to the memories and interpretations offered by veterans of both campaigns. Veterans commonly asserted that despite the

French withdrawal from Algeria, there had at least been 'no Dien Bien Phu' there.[42] This shifted the focus onto a political rather than a military defeat, and thus perpetuated the collective memory of Dien Bien Phu which had emerged in the years following defeat, ensuring that this memory would be maintained in the post-colonial era. In the light of the Algerian War, Dien Bien Phu might have seemed even more serious – the beginning of a long process of defeat – but it was also easier to explain, being a military defeat compounded by political incompetence, whereas for some veterans the loss of Algeria did not even represent a military defeat.

Political uses of the memory of Dien Bien Phu after the end of empire

After 1962, the American War in Vietnam allowed those in France who were determined to keep alive a memory of Dien Bien Phu to share this burden of remembrance and interpretation. As Benjamin Stora argues,

> Dien Bien Phu . . . the first real military crisis for the 'white man' for a long time, is a pitiful spectacle which they no longer have to bear alone. The Indochina War paid the price for the emergence of this substitution in the imagination . . . this dual presence and absence of Indochina is decisive.[43]

Yet, if the American War in Vietnam proved 'decisive' in keeping a certain French memory of Dien Bien Phu alive into the 1970s, this also further facilitated the politicization of veterans' and politicians' memories; under the influence of the Cold War rhetoric used to justify the American presence in Vietnam, the French defeat took on a new political and ideological aspect in the post-colonial era. The key point in the post-Algerian War period, in respect of commemoration, remembrance or interpretation of Dien Bien Phu's significance, was the fall of Saigon in 1975. This galvanized the French Right – the old 'colonial party' – whose collective memory of Dien Bien Phu as a disaster, the beginning of an ineluctable decline, a betrayal both of a good cause and of loyal natives, was revived: as one veteran argued, 'It was at Dien Bien Phu that the Americans, too, began to lose their war, not understanding that this defeat for France was also that of the West.'[44]

New interpretations, from 1975 in the aftermath of the fall of Saigon, of what Dien Bien Phu signified both expressed the prevailing collective memory on the Right of the loss of Indochina and added a further layer to it, that of a presumed decline of the West as a result of the failings first of France and now the United States. For Yves Gignac, president of the Indochina veterans' association – which had been very active in the attempts to overthrow the Fourth Republic, seen as responsible for the defeat, in 1954 and 1958[45] – with the fall of Saigon France 'had died for the second time'. The publisher and writer Philippe Héduy, known for his support of *Algérie française* rather than any personal attachment to Indochina, reflected in 1975 that Dien Bien Phu had appeared as the beginning of an inevitable decline:

> In 1954, the May bell had tolled for a country which was ours, and what would follow for us, in an obvious sequence, would come inexorably : we knew it. We knew, but that was no reason not to oppose our destiny. It was a question of dignity.[46]

None of Héduy's activities in favour of French Algeria between 1954 and 1962 gave the impression of a man supporting a lost cause; for this individual with no personal experience of Dien Bien Phu but an obvious intellectual and emotional link to the Indochina War via his attempts to keep Algeria French, the 'memory' of Dien Bien Phu now fitted into a longer narrative of French failures.

Nostalgia for the days of colonial rule is a recurring feature of memories of Indochina from the period around the twentieth anniversary of Dien Bien Phu and the fall of Saigon. For Gignac, for example, 'Today, with the loss of the freedoms which our country had brought the people of Vietnam, Cambodia and Laos, France is leaving a second time.'[47] The memory of the French colonial 'mission' had become bound up, after the withdrawal from Algeria, with that of Dien Bien Phu, and the American defeat served to reawaken this nostalgia and regret. Gaullist deputy and campaigner for French Algeria, Raymond Dronne, further developed the theme of a second defeat for France in Vietnam:

> The bell has tolled a third and final time, after Dien Bien Phu and Geneva, for what was French Indochina . . . It's the funeral bell of a civilization which has rung out. In this French Indochina an original culture was born and took root, a synthesis of Western, humanist culture in the French style and of the old, Chinese-style far-eastern culture . . . There, between 'colonized' and 'colonizers' there was no wall of religion like in the Muslim countries of North Africa nor a gulf between the civilized, 'evolved' peoples and the still primitive humanity like in Black Africa[48]

It is perhaps no surprise to hear Right-wing deputies appropriate the memory of a colonial defeat for the cause of colonial nostalgia. However, the dimensions which the French Right's collective memory of Dien Bien Phu had acquired by 1975 do suggest a memory of defeat or betrayal beginning to acquire an ideological aspect that was absent from earlier reflections on the battle's significance. For General Edmond Jouhaud – convicted for his part in the failed 1961 putsch – the abiding lesson of Dien Bien Phu was that 'The spirit of subversion, of resignation, triumphed over honour. Can the West arrest this dangerous descent which is leading to its loss?'[49] Jacques Soustelle, the former Gaullist intellectual whose commitment to French Algeria led him into exile on account of his involvement with the military and settler revolts in the colony, despite having remained virtually silent at the time of Dien Bien Phu, proclaimed 20 years later that 'For our civilization, the loss of Indochina has the same meaning as that of Dacia and Great Britain for Rome. It's the ebb.'[50] By 1975, therefore, the memory of Dien Bien Phu on the French nationalist and nostalgic Right had become one of the decisive stage in an apocalyptic battle of civilizations; the issues of heroism, betrayal and political incompetence had not vanished but the memory of the details of the defeat had receded, allowing another version of events, one which accorded much greater long-term vision to the critics of the French withdrawal from Indochina, to come to prominence.

The memories of Dien Bien Phu discussed here have focused on a fairly well-defined and distinctive group: those elements of the French military and political classes who had cause to draw upon their shared and evolving memories of the French defeat in Indochina for the purposes of their opposition first to decolonization in Algeria and then to what they saw as a more general decline in French and Western prestige. Yet there is little evidence of a more generalized, less politicized alternative memory of Dien Bien Phu; like the war in Algeria, the Indochina campaign remained largely absent from French national memory through the 1970s and 80s. From 1975, the *Association nationale des combattants de Dien Bien Phu* (ANCDBP) was recognized as the main veterans' body, and has held regular commemorations, including a large event in Pau every five years. In 1989 the ANCDBP called for the government to pay reparations to former prisoners of war from the Indochina War. Three years earlier, it had been decided to construct an Indochina War memorial, at Fréjus. Nevertheless when President François Mitterrand visited Dien Bien Phu before inaugurating the memorial, in 1993, veterans' associations protested about the order of the two commemorations and some veterans' leaders refused

Mitterrand's invitation to attend the ceremony. Only one ex-parachutist among those interviewed for a recent thesis used the word 'defeat' to describe Dien Bien Phu.[51]

The collective memory, among certain sections of the French Right, of the French defeat in Indochina in 1954 can, therefore, be seen as an example of how the context in which memories took shape had a decisive impact in determining which aspects of the defeat were remembered. Veterans' memories of their own experience quickly fused with the conclusions that were drawn in the immediate aftermath of Dien Bien Phu, so that by the time of the war in Algeria, what was remembered was not only the defeat but also the veterans' impression of its causes and consequences. The persistence of the impressions formed, in response to the French political climate and the war in Algeria, in the years following the defeat, suggests that in the case of Dien Bien Phu, the context in which this defeat was remembered facilitated its transformation from personal humiliation to national disaster.

Notes

1. Raoul Salan, *Mémoires: fin d'un empire. Vol. 2: 'Le Vietminh mon adversaire' octobre 1946-octobre 1954* (Paris: Presses de la Cité, 1971), p. 442. Salan was commander-in-chief of the French expeditionary corps in Indochina between January 1952 and May 1953.
2. A notable exception to the relative lack of scholarly interest in memories of Dien Bien Phu is Alain Ruscio and Serge Tignères, *Dien Bien Phu, mythes et réalités 1954–2004: cinquante and de passions françaises* (Paris: Les Indes Savantes, 2005).
3. On questions of commemoration in post-war France, see Gérard Namer, *La Commémoration en France de 1945 à nos jours* (Paris: L'Harmattan, 1987).
4. Pierre Messmer, 'Préface' in Alain Griotteray, *Dien Bien Phu, pourquoi en est-on arrivé là?* (Monaco: Editions du Rocher, 2004), p. 10.
5. Griotteray, *Dien Bien Phu, pourquoi en est-on arrivé là?*, p. 135.
6. On the memory of 1940 as a 'debacle' and its influence on French reactions to the wars of decolonization, see Julian Jackson, *The Fall of France: The Nazi Invasion of 1940*, (Oxford: Oxford University Press, 2003), pp. 245–7.
7. Paul Boury, *Dien Bien Phu: pourquoi? Comment? Et après?* (Dijon: Cléa, 2004), p. 272.
8. Roger Bruge, *Les hommes de Dien Bien Phu* (Paris: Perrin, 2003), p. 607.
9. Jean-Yves Alquier, 'Historique: Vietnam 1940–1975', in Alquier (ed.), *Chant funèbre pour Pnom Penh et Saigon* (Paris: S.P.L., 1975), p. 395.
10. Raoul Girardet, *Singulièrement Libre* (Paris: Perin, 1990), p. 133. Violent clashes between nationalist and Communist students in Paris were not entirely uncommon in 1950s Paris; on the political context, see Olivier Dard, 'Jalons pour une histoire des étudiants nationalistes sous la IVè République', *Historiens et Géographes* no. 358 (1997), 249–63.

11. Jean-Marie le Pen, 'Le message de Léonidas', in Alquier (ed.), *Chant funèbre*, p. 212.
12. On the notion of 'social frameworks', see Maurice Halbwachs, *On Collective Memory*, trans. by Lewis A. Coser (Chicago: University of Chicago Press, 1992).
13. Roger Bruge, *Les hommes de Dien Bien Phu* (Paris: Perrin, 2003), p. 31.
14. *France-Soir*, 9 May 1954.
15. *Le Figaro*, 8 May 1954.
16. 'La leçon de Dien Bien Phu', *Paris-Match*, 12 May 1956.
17. *L'Aurore*, 10 May 1954. Emphasis in original.
18. This was especially true in 1957–8, when fears of internationalization of the Algerian War prompted Gaullists to rally round the cry of 'no diplomatic Dien Bien Phu'.
19. Charles de Gaulle, *Discours et Messages, Vol. 2: Dans l'attente, février 1946-avril 1958* (Paris: Plon, 1970), pp. 617–18.
20. Charles-Robert Ageron has demonstrated that public opinion did not see the war in Indochina as a particularly high priority and that a majority supported negotiation or withdrawal even before Dien Bien Phu. See C. R. Ageron, 'L'Opinion publique face aux problèmes de l'Union française' in Institut d'Histoire du Temps Présent, *Les chemins de la décolonisation de l'empire français 1936–1956* (Paris: CNRS, 1986), pp. 33–48. On the May 1954 demonstration, see Frédéric Turpin, 'Printemps 1954. Echec à de Gaulle: un retour au pouvoir manqué', *Revue Historique*, no. 306 (2004), 913–27.
21. Jacques Soustelle, speech in National Assembly, 9 June 1954, *Journal Officiel de la République Française: Débats de l'Assemblée Nationale.*
22. *Le Figaro*, 8 May 1954.
23. On the strained relations between France and the United States over the war in Indochina and Dien Bien Phu, see Lawrence S. Kaplan, Denise Artaud, and Mark Rubin (eds), *Dien Bien Phu and the Crisis of Franco-American Relations, 1954–1955* (Wilmington, DE: SR Books, 1990).
24. The final battle at Dien Bien Phu was known as Operation Castor.
25. Jean-Marie Juteau, quoted in Journoud & Tertrais, *Paroles de Dien Bien Phu*, pp. 348–9.
26. Michel Bodin, *Les combattants français face à la guerre d'Indochine: 1945–1954* (Paris: L'Harmattan, 1998), p. 257.
27. Ibid., p. 247.
28. Ibid., p. 242. The French Empire had been renamed the French Union in 1946.
29. Ibid., p. 253.
30. Ibid., p. 254.
31. Henri Navarre, quoted in Benjamin Stora, *Imaginaires de guerre: Algérie-Vietnam, en France et aux Etats-Unis* (Paris: La Découverte, 1997), p. 96.
32. Navarre, from *Le temps des vérités*, quoted in Anne Logeay, 'Qui se souvient de Dien Bien Phu? Quelques remarques sur l'érosion de la mémoire' in Société Française d'Histoire d'Outre-Mer, *1954–2004: la bataille de Dien Bien Phu entre histoire et mémoire* (2004), p. 183.
33. On the French army's doctrine of revolutionary warfare, based on its experiences in Indochina, see Peter Paret, *French Revolutionary Warfare from Indochina to Algeria: The Analysis of a Political and Military Doctrine* (London: Pall Mall Press, 1964).

34. For a good discussion of the effects of the defeat in Indochina on the mentality of French officers in Algeria, see Martin Alexander, 'Seeking France's "Lost Soldiers": Reflections on the French military crisis in Algeria', in Kenneth Mouré and Martin Alexander (eds), *Crisis and Renewal in France 1918–1962* (Oxford: Berghahn, 2001).
35. Logeay, 'Qui se souvient de Dien Bien Phu?', pp. 184–5; on the High Command's concern about officers' tendency to draw comparisons between Indochina and Algeria, see also Raphaëlle Branche, *La torture et l'armée pendant la guerre d'Algérie, 1954–1962* (Paris: Gallimard, 2001), p. 40.
36. Logeay, 'Qui se souvient de Dien Bien Phu?', p. 186.
37. Jean Pouget, quoted in Stora, *Imaginaires de guerre*, p. 97.
38. Ibid., p. 254.
39. Jean Pouget, *Nous étions à Dien Bien Phu* (Paris: Presses de la Cité, 1964).
40. Hélie Denoix de Saint Marc quoted in Valérie Padilla, 'Références indochinoises et problèmes algériens', *Histoire et Défense. Les Cahiers de Montpellier*, vol. 23, no. 1 (1991), 70–1.
41. Hélie Denoix de Saint-Marc, quoted in Padilla, 'Références indochinoises', p. 71.
42. Stora, *Imaginaires de guerre*, p. 96.
43. Ibid., pp. 13–14.
44. Jean Lartéguy, *L'adieu à Saigon* (Paris: Presses de la Cité, 1975), p. 21.
45. On the role played by veterans' movements in the Gaullists' attempts to bring down the Fourth Republic, see Christophe Nick, *Résurrection: naissance de la Vè République, un coup d'état démocratique* (Paris: Fayard, 1998).
46. Philippe Héduy, 'Le grand souffle apaisant', in Alquier (ed.), *Chant funèbre*, p. 158.
47. Yves Gignac, in Alquier (ed.), p. 142.
48. Raymond Dronne, 'Mort d'une civilisation', in Alquier (ed.), pp. 105–6.
49. Edmond Jouhaud, 'Les horizons parsemés d'exigences', in Alquier (ed.), pp. 181–2.
50. Jacques Soustelle, 'Le reflux de l'occident', in Alquier (ed.), pp. 322–4.
51. Hugo Gérin, *La mémoire des parachutistes à Dien Bien Phu: 13 mars – 7 mai 1954: entretiens et récit anthropologique* (Panazol: Lavauzelle, 2004), p. 197.

14

The Enduring Paradigm of the 'Lost Cause': Defeat in Vietnam, the Stab-in-the-Back Legend, and the Construction of a Myth

Jeffrey Kimball

In his influential tour de force, *The Culture of Defeat: On National Trauma, Mourning, and Recovery* (2001), Wolfgang Schivelbusch observed that 'the absence of a stab-in-the-back legend in the United States after the Vietnam War shows that accusations of betrayal can grow only if the political and historical soil in which they are planted is fertile'.[1] In an endnote to this sentence, Schivelbusch incongruously cited an article I had published 13 years earlier entitled 'The Stab-in-the-Back Legend and the Vietnam War', in which I had arrived at conclusions incompatible with his. I had argued that American policy defeat in the Vietnam War had indeed produced a powerful myth of betrayal that was analogous to the archetypal *Dolchstoss* legend of post-World War I Germany.[2] Originating during the Vietnam War in the bitter debate between Americans over US policy and strategy – I had written – the stab-in-the-back myth developed into a full-fledged explanation for American defeat after the war ended and as a related debate unfolded over the causes of failure and the future of policy. The myth, or legend, blamed leftists, liberals, the press, the antiwar movement, civilian policymakers, Democratic Party presidents, and the Congress of the United States – and particularly the 'dovish' representatives within it – for snatching defeat from the jaws of victory. It included the circumstances of defeat in war, consequent national humiliation, and the scapegoating of others by war hawks on the Right. Soon, it evolved into a larger myth of the Lost Cause of the Vietnam War.

Schivelbusch acknowledged that the stab-in-the-back rhetoric I had documented was 'quite similar to that voiced by right-wing Germans during the Weimar Republic', and that it raised 'the question of whether the stab-in-the-back legend was a uniquely German phenomenon', which may 'have had a similar effect in the United States'. Nonetheless he maintained that American soil was not fertile ground for the emergence of a *Dolchstoss* legend, because the Vietnam War 'did not entail national collapse, . . . was not followed by a humiliation like that of the Versailles Treaty, . . . [and] did not polarize the nation or lead to civil war'. Furthermore, he asserted, the United States did not possess a requisite historical mythology of betrayal; that is, a mythological paradigm similar to that provided by the novels of Sir Walter Scott for the post-Civil War American South, *La Chanson de Roland* for post-Franco-Prussian War France, and the *Nibelungenlied* for post-World War I Germany.

Schivelbusch added that the Vietnam War's impact on the US did not meet two other important criteria. First, he intimated that accusations of betrayal must be socially or politically 'counterrevolutionary', by which he apparently meant that stab-in-the-back scapegoating must extend beyond the assignment of blame to individuals and must instead be directed against a collectivity and bring about the overthrow of revolutionary groups or their reforms. Second, Schivelbusch questioned whether the ending of the Vietnam War could legitimately be considered a 'true defeat or simply the first war the United States had not won conclusively'.[3]

Schivelbusch, I submit, was incorrect on virtually every count.[4] Contrary to his claims, there was indeed fertile ground in the United States at the time of the Vietnam War for a stab-in-the-back legend to emerge. To begin with, there did exist a pre-existing historical mythology of betrayal in defeat. In the distant past, War Hawks during the War of 1812 had accused New England Federalists of near-treasonous activity for their roles in the failure of American arms to win all of its battles and to conquer Canada. They had also charged Congress and the president of naïveté, confusion, and parsimony and the non-professional citizen-soldier militia of ineptitude. In the aftermath of the Civil War, unrepentant Confederate rebels had explained the defeat of the Confederacy as having been caused by President Jefferson Davis's incompetent interference in military strategy and the mistakes of General Robert E. Lee's subordinates, while some later historians had found fault with the states' rights sentiment of Southern politicians and the excessively democratic and antimilitary attitudes of the Southern white masses. During the Korean War a century later, the Right and much of the Republican Party had castigated President Harry S. Truman and Secretary of State Dean Acheson for their alleged military unpreparedness and exclusion of South Korea from a Pacific defence perimeter, which, they argued, had encouraged North Korean aggression. With Truman's failure to 'liberate' North Korea, the Right had also reproached him for his rejection of

General Douglas MacArthur's recommendations to bomb China. These charges were of the same genre as Right and Republican claims that the Franklin D. Roosevelt administration had sold out non-Communist Poland and Nationalist China in the Yalta Agreement of February 1945, and that the Truman administration had allowed China to be 'lost' to the Communists in the aftermath of World War II.

These examples suggest that Americans were and are the heir of the betrayal myths or mentalities of 'Western civilization' dating from the Iliad to the present. Moreover, the reappearance of the stab-in-the-back theme in connection with the Vietnam War is but a repetition of a familiar pattern of response to defeat in all societies, whether Western or non-Western, a pattern in which defeat produces recrimination. Details may differ from culture to culture and epoch to epoch, but the essential characteristics remain the same. Beyond historical fallacy, cultural mythology, and misplaced pride, stab-in-the-back accusations reflect tensions between civilians and the military within democratic republics and the outlooks and interests of specific class and ideological groups in all societies – especially the psychological-cultural-class tendency of militaristic rightist, traditional, anti-democratic, or conservative elements to explain defeat by accusing targeted 'others' of betrayal.

Also contrary to Schivelbusch's assertion, Americans were indeed polarized by the war. By the late 1960s, Americans on the home front were sharply divided about whether to support their government or whether the war was unwise, immoral, and too costly. They were perplexed and divided on whether their government had lied to them about the cause of the war, whether or not to protest government policy and strategy, whether protest during wartime was patriotic, and whether the United States should escalate the war or withdraw from Vietnam. If withdrawal was the issue, Americans disagreed about how to withdraw and how rapidly. Public opinion polls revealed consistent disagreement, for example, between those who favoured some sort of American withdrawal (from immediate to gradual) and those who favoured staying the course or escalating drastically (from keeping military pressure on the enemy to using nuclear weapons). In September 1968, 42 per cent of those polled favoured one of the options in the first category, 43 per cent favoured one of the choices in the second, and 9 per cent registered 'no opinion'.[5]

The aggregate 'public' was actually many 'publics' distinguished by region, religion, race, class, education, gender, age, political party affiliation, labour union membership, and other viable group categories. Polls did not measure group differences consistently or precisely, but there was plenty of evidence that differences existed and that these reflected deep divisions in American society. In addition, 'followers' tended to rally behind presidents during crises and to change their opinions on cue from their leaders. In April 1968, after President Lyndon B. Johnson's announcement of a bombing halt, 64 per cent of those polled approved the decision; but in March, before the announcement, only 40 per cent had voiced support for stopping the bombing.[6]

In addition to these sociological factors, there were situational factors, such as the war's costs and frustrations, that also drove public opinion. The most palpable were mounting casualties, feelings of futility vis-à-vis prospects of victory in Vietnam, increasing economic burdens related to the spiralling costs of war, social dissonance growing out of civil rights battles, antiwar protests, and counterculture challenges. Since the introduction of combat troops in 1965, US battle casualties had risen slowly in contrast to the Korean War, in which the numbers of killed and wounded had been greater in the first year of fighting. But the cumulative total of US casualties in Vietnam had exceeded the casualty total of the entire Korean War by the middle of 1968, which was also the year of peak per-annum combat casualties in Vietnam. This crisis year also brought more inflation; new taxes; and difficult choices between expending money for waging war, maintaining America's global military force, and domestic programmes. Other social fissures – racial division, the alienation of the blue-collar and middle classes from their political parties and government, and youthful 'counterculture' challenges to 'mainstream' values – widened with the war. It was a war that was already too long, and after the Tet Offensive commenced on 30 January 1968, it appeared to be irrevocably deadlocked, with no end in sight.[7]

The public, and particularly the World War II generation, was perplexed about the causes of stalemate. How could this small, underdeveloped country of Vietnam stymie the United States, which had recently defeated Germany and Japan and contained the Soviet Union? Many Americans looked inward towards America's own divisions, presidential mismanagement, and ineffective strategies, and with this introspection an incipient stab-in-the-back myth emerged on the Right. Hawks blamed the allegedly restrained military strategy of President Johnson and his liberal civilian advisers for the impasse. From politicians such as Barry Goldwater – 'Mr Conservative' – to right-wing media to hard-line generals and admirals in the Joint Chiefs of Staff, the call of hawks had been for more troops, more bombing, a naval blockade, and ground attacks against North Vietnam, and even the use of nuclear weapons, regardless of the risk of war with China or the Soviet Union.[8] They argued that the pace of the administration's military escalations since 1964 had been too gradual, thereby allowing the enemy to adapt, prepare, and resist. Had more American troops been deployed in South Vietnam more rapidly, had more massive air power been thrown against more vital targets much sooner, the North Vietnamese and Vietcong would have been forced to capitulate or would have made the requisite diplomatic concessions. Hawks also blamed the dovish antiwar opposition-the 'other' within-for having contributed to the undermining of America's will to win. The antiwar movement, they maintained, had caused disunity at home, encouraged the enemy to persist, and discouraged American leaders from prosecuting the war more forcefully. They complained that the movement was assisted by the press, particularly the television press, which, they asserted, excessively reported both the horrors of the war in Vietnam and the protests against it at home. Conservative leaders such as Henry Kissinger believed that social and political conditions in the United States approximated those of 'civil war'.[9]

Wartime fissures manifested themselves dramatically in the realm of politics. In the presidential campaign of 1968, for example, each of the two major political parties was fractured along conservative, 'moderate', and liberal lines regarding the keystone issues of the Vietnam War, the Cold War, and social reform at home. These fractures reflected divisions in the population at large and the melding of international and domestic concerns. On the sidelines, Governor George Wallace of Alabama ran an independent, populist, segregationist, prowar campaign focused on wooing white Southern, rural, and labour votes. For his part, Richard M. Nixon, the leading Republican candidate, fashioned a 'peace-with-honor' rhetorical campaign strategy, wherein his calls for negotiations and Vietnamization appealed to moderates and centrists who wanted a negotiated settlement in Vietnam and détente with the Soviet Union, while his emphasis on honour, strength, and commitment simultaneously appealed to his hawkish, right-wing base.[10]

Disconcerted by his inability to conclude the war satisfactorily and challenged from within his own party, President Johnson withdrew from presidential race in March. It was only one of many political shocks that year. The visible events included the assassinations of Martin Luther King and Robert F. Kennedy, violent spontaneous uprisings (aka riots) in the black ghettoes of major cities, the Chicago police riot (aka violent suppression of dissent) during the Democratic Party convention, and major shifts in political voting patterns. The 'invisible' events included growing dissent early in 1968 from President Johnson's militant strategy towards the prosecution of the war among former Cold Warriors within

Johnson's own administration, the secret steps taken later in the year by candidate Richard Nixon in league with President Nguyen Van Thieu of South Vietnam to sabotage Johnson's halting steps towards negotiations with the Communists, and the growing strength of a far-right political movement, which had begun organizing in earnest after Barry Goldwater's defeat in the presidential election of 1964.[11]

By the time the war ended, America's adversaries and most of its own citizenry *perceived* the nation to have been defeated; that is, they believed their nation, or at least its government, not only failed in achieving its stated policy goal of turning South Vietnam into an independent state in the name of anticommunism but also failed in vanquishing, conquering, or subduing the military and political forces of the enemy. These perceptions were palpable. American national policy and military strategy had met with defeat. It was a true defeat – not simply, as Schivelbusch claimed, a war the United States failed to win conclusively. In the end, US armed forces withdrew and the other side reunified Vietnam under a Communist government – a result the United States had vigorously and violently opposed for 30 years.

There were other humiliating defeats. Even though American armed forces in Indochina were not beaten in battle *en masse*, for example, small infantry units – squads and platoons – lost skirmishes and battles.[12] For this and other reasons, by the early 1970s the morale of the entire US Army was in shambles. In Vietnam, there were numerous incidents of rank-and-file mutiny, combat refusal, and 'fragging'.[13] In America draft-age men resisted conscription in various ways, including flight to Canada and elsewhere, and soldiers deserted in large numbers. Many of the soldiers, airmen, and sailors who served in Vietnam returned home frustrated and angry – some directing their ire against the succession of governments that sent them to Vietnam and the officers who commanded them, others directing their anger against those who opposed the war at home. In both cases, especially among those who engaged in combat, many felt abandoned, unappreciated, and scorned. Criticized from within and without for its failures of strategy, administration, and leadership, the military's high command was humiliated.

The reality and perception of American defeat in Vietnam symbolically overturned the myth of American military omnipotence and the tradition of invincibility. While American television audiences watched news of a People's Liberation Armed Forces' tank crashing through the gate of the Presidential Palace in Saigon on 30 April 1975, there seemed little doubt that the decades-old American policy of intervention in

Vietnam had been defeated. There were other poignant, memorable images of defeat: panic-stricken soldiers of the Republic of Vietnam mobbing evacuation aircraft on take-off, 'Huey' helicopters lifting evacuees from the precarious perch of the United States Embassy's rooftop helipad, and US Navy personnel shoving an empty evacuation helicopter into the emerald green South China Sea from the overcrowded deck of an American aircraft carrier. For many Americans, it was a humiliating consequence.

By the time the war ended, two presidencies, those of Johnson and Nixon, had fallen victim to the war. At the same time, the Democratic Party underwent an upheaval in leadership and organizational structure, opening a deep chasm between civil rights, antiwar, centrist, and left-leaning reformers, on the one hand, and prowar, conservative-leaning Cold Warriors, old guard politicos, union officials, and Southern segregationists, on the other hand. Wounded by Nixon's Watergate scandal, itself a by-product of the Vietnam War, the Republican Party was initially plunged into an even smaller minority status – only to recover in the Thermidorian reaction of the 1980s (aka 'the Reagan Revolution') and the counter-revolution of the 1990s (for example, Republican Party congressional victories in 1994). These gains extended into the early twenty-first century with the presidency of George W. Bush, himself a baby-boomer (b. July 1946) who had come of age during the Vietnam War and the 'long sixties' – the period from approximately 1956 to 1975.[14]

During the 1960s and 1970s, both major parties experienced significant and long-lasting shifts in their base constituencies, which in the long term was no less than revolutionary, albeit more or less non-violent. The white South, having been solidly Democratic since before the Civil War, became solidly Republican, while Catholic, ethnic, and rank-and-file labour voters divided their support between the parties on the basis of their attitudes towards 'patriotic' and 'cultural' issues associated with the Vietnam War and Great Society social reform, as well as cultural issues that dated from the 1960s.

From the late 1960s into the 1970s, moreover, the US economy was under increasing strain (in part because of the costs of the Vietnam War). Beginning with the Vietnam War engendered 'stagflation' and the OPEC-initiated oil crisis of the Nixon years, there followed a post-war period of economic crisis in the form of inflation and trade imbalances, accompanied by apparent policy confusion concerning monetary, foreign, and domestic policy during the presidencies of Gerald R. Ford and Jimmy Carter. In particular, American foreign and military policy doctrines were in tatters, notably the Cold War consensus concerning military intervention in 'Third World' revolutions. Then, in the presidential election of 1980, the rightist counter-revolution of Ronald Reagan achieved power through allusions to betrayal in Vietnam and calls for the revival of patriotic pride, support for pro-interventionary and hard-line foreign and military policies, increases in military spending, the re-establishment of 'law and order' at home, and adherence to 'traditional moral values'.[15]

Although Schivelbusch identified individuals and social groups as agents of stab-in-the-back legends and Lost Cause crusades, he privileged culture as the prime cause.[16] One of the major difficulties with a cultural explanation of the stab-in-the-back phenomenon is that for each nation in which such a legend comes into existence only part of that culture's population subscribes to it, while at least one other part of the society is the target of accusation. It seems to me that greater attention should be placed on the subculture, mentality, and motives of the accusing group and its followers, as well as on the methods that the accusing group uses to create such a legend and persuade a critical mass of others to accept it. I have argued that during and after the Vietnam War, militaristic, conservative, and right-wing individuals and groups, including other segments of the population – for example, a significant proportion of white Southerners, military veterans, Christian evangelicals, opinion leaders, and hawkish policymakers and military commanders – were the staunchest proponents of and believers in the stab-in-the-back legend for reasons having to do with their mindset, vested interests, cynical opportunism, and other such motives and ideological-psychological predispositions. There has been no lack of evidence in the form of opinion polls, public statements, memoirs, newspaper accounts, op-ed columns, and histories to support this observation.

During the past ten years, furthermore, considerably more archival evidence from the files of President Nixon's White House has become available, throwing more light on how elite individuals and groups legitimated stab-in-the back scapegoating. These archives demonstrate, for example, not only how Nixon and his aide Henry Kissinger deceived the public about their exit strategy but also how they set out to shape Americans' memory of how and why the United States suffered defeat in Vietnam.

The new documents reveal that it was Nixon and Kissinger and not the antiwar movement who had prolonged the war and the ceasefire negotiations so that a settlement would not be reached until the 1972 US presidential elections. The purpose of this timing was to ensure that if the Saigon government fell, the fall, coming sometime after the election, would not destroy Nixon's chances for a second presidential term. In addition, the prolongation of the war would also enable Nixon and Kissinger to further strengthen Saigon's forces and weaken Hanoi's, so that if the Saigon regime collapsed, it would likely do so after a sufficiently long interval had passed following the diplomatic settlement; hence, their policies would not appear to have been responsible for defeat. They could hope that neither the American public nor America's allies or adversaries would therefore perceive Saigon's fall as a humiliating US defeat. Unknown to the public, both hawks in Congress and Nixon in the White House had come to believe that it would be impossible to win the war in any imaginable cost-effective way, that, in other words, the war was a Lost Cause. But even before the ceasefire agreement of late January 1973, Nixon and his allies were prepared to place blame for the defeat on opponents of the war.[17]

After his re-election in November 1972, and as he was preparing to sign a ceasefire agreement, Nixon set out to shape public perceptions of his role in ending the American phase of the Vietnam War. His plan was to mount a broad, concerted public relations campaign that would portray him as both a peacemaker and tough-minded war leader. By going on the attack, he could not only counter antiwar charges that he had prolonged the war but he could put the albatross of responsibility for defeat around the neck of the antiwar opposition and bolster his image among hawks. Nixon's self-serving public-relations campaign, which he carried out well into the 1980s through popular memoirs and histories, provided the critical legitimizing element for a stab-in-the-back narrative of the termination of the war, whose origins could be traced to earlier right-wing criticisms of the antiwar movement, liberal policymakers, Congress, and the press.

Another key element in the emergence and growth of a stab-in-the-back legend was Nixon's role in manufacturing the POW/MIA myth. Early in his administration, Nixon had publicly announced that one of his major policy purposes in continuing the war was to win the release of American prisoners of war (POWs). But his real purpose in making POW releases into a patriotic political issue had been to counter the antiwar opposition's critique of the war as immoral, cost ineffective, and unwise, thereby building sufficient public support for his continuance and prolongation of the war.[18]

By the time of the agreement in January 1973, the POW question had merged with Americans' concern about US military personnel missing in action (MIAs). For the next two decades, many continued to believe that POWs *and* MIAs were still held prisoner in Vietnam. It was a falsehood fostered by fortune seekers and right-wing politicians, which, in early 1980s, morphed into a full-blown myth, as portrayed in popular movies like *Rambo: First Blood Part II* (1985). *Rambo* was nothing less than an allegory of the alleged betrayal of American soldiers during the Vietnam War, a war they had not been allowed to win and which had ended with a peace that had left many behind. President Ronald Reagan, who loved the film,[19] exploited this enlarged version of the stab-in-the-back legend for political purposes.

The experience of fighting and arguing about a protracted and ultimately failed war had the effect of shattering the 1950s–1960s national consensus in the United States in support of military intervention in anti-colonial revolutions. The Vietnam War now replaced World War II as *the* historical 'lesson' that informed and justified American foreign policy – the script, the narrative to follow. In the wake of defeat in Vietnam and social conflict at home, however, American foreign policy policymakers drew different historical lessons and wrote different narratives. These in turn evolved into conflicting paradigms that defined and shaped foreign policy perspectives and policies in the decades to come.

Liberal and pacifist doves believed that intervention in the 'quagmire' of Vietnam was a monumental misadventure to be conscientiously shunned in analogous Third World trouble spots. Underpinning this 'Vietnam Analogy' were several assumptions: US involvement in Vietnam had been an avoidable mistake, militarism and the national security state had helped to cause the disastrous intervention, Americans' ability to influence and even understand other cultures was limited.[20]

For their part, hawks on the right of centre rejected both the Vietnam analogy of doves and the anti-imperialist analysis of the Left. To hawks, the Vietnam analogy was nothing less than a 'Vietnam Syndrome', a dovish symptom of psychological weakness that undermined America's will and ability to police the globe and intervene in anti-capitalist revolutions and crises, thus eroding its great-power credibility. Hoping to stamp out this 'isolationist' aberration, and differing with Europe-first, anti-Soviet Cold Warriors by their greater concern with the Third World, they promoted the view that military intervention in developing countries should be seen as a necessary and viable tool of US foreign policy. The experience of the Vietnam War, they insisted, had demonstrated that revolutions can be crushed – if only the correct strategy were pursued.

Joining their ranks were former liberal and leftist supporters of Israel, now 'neoconservatives', who had grown concerned that the anti-Vietnam War and the anti-Cold War Left's sympathy with Palestinian claims of unjust treatment at the hands of Israelis would weaken revolutionary containment and endanger the future survival of Israel. Despite the Right's history of anti-Semitism, neoconservatives' concern for the fate of Israel led them to feel increasingly more comfortable with the Right's staunch support of anti-leftist containment policies.[21]

Conservative and neoconservative revisionists not only claimed that the Vietnam War had been moral and idealistic – in stark contrast to the dovish charges that the means America used in fighting the Vietnam War had been immoral and its purpose self-aggrandizing – but that it could have been won had civilian policymakers pursued near total-war strategies, stood by South Vietnamese allies, and allowed the US military to fight without constraints. Revisionists further asserted that the confusion characterizing civilian leadership in the presidency and Congress would not have appeared if antiwar demonstrators, students, professors, movie stars, and the electronic press had not undermined the will of the American people by their activities and portrayals of the fighting. Homefront criticism and turmoil, revisionists maintained, had a particularly adverse effect on the army's efficiency, and, in addition, encouraged the enemy to persevere. The revisionist/hawk argument had several versions, some more strident, some more subtle than others. Yet the suggestion in all was that the war had been winnable but the military had been figuratively stabbed in its back. Defeat had consequently been snatched from the jaws of victory.[22]

Reeling from other setbacks in Angola, Mozambique, Ethiopia, Grenada, Iran, and Nicaragua, and troubled by the public's aversion to intervention in less-developed countries, some hawkish political leaders in the post-war era actively sought to overcome the Vietnam Syndrome. President Ronald Reagan, for example, used bellicose rhetoric to reinvigorate the public's Cold War mentality. Time and again, he drew on revisionist explanations of defeat to evoke the memory of the noble Lost Cause of the Vietnam War and of the POWs and MIAs allegedly left behind, promising that never again would the United States send its boys to a war: 'our government is afraid to let them win.' Thus, he acknowledged the public's concern about future quagmires, but he simultaneously reinforced its habitual anticommunism, affirmed a commitment to interventionism, and suggested that his own interventions would be quick, forceful, and decisive.[23]

Although partially successful in remilitarizing the electorate and reinvigorating the morale of interventionist national security managers, Reagan's threats of US military intervention were still not followed by massive, Vietnam-like invasions of developing countries. Besides the public's post-Vietnam aversion to large-scale intervention, interventionists faced another important obstacle: bureaucratic caution in the Pentagon. Flowing out of their analysis of defeat in Vietnam, Pentagon strategists' caution was not so much a rejection of intervention as it was a mark of bureaucratic fear of being held responsible for possible defeat. Many in the Pentagon, including General Colin Powell, had come to believe that prerequisites for success in sending American troops into situations resembling Vietnam included the public's wholehearted support, the authority to use whatever force the military determined was necessary to win, and perhaps even a declaration of war.

The US government applied some of these lessons during the Gulf War of 1991 in the form of greater press censorship and the early application of overwhelming firepower on the ground and in the air. Although President George H. Bush stopped short of occupying Iraq for fear of sparking another Vietnam-like guerrilla war and dealing with the difficulties of state building, he nonetheless declared, 'By God, we've kicked the Vietnam Syndrome once and for all.' Of course, he had not, and in fact he had avoided removing Saddam Hussein from power fearing that an occupation of Iraq would become a Vietnam-like quagmire.

His son, a born-again Christian Southerner, who had evaded military service in Vietnam, had worked during the Vietnam War for the senatorial campaign of a conservative Alabama candidate – a campaign that, among other things, accused antiwar doves of stabbing the military in its back. Coming to presidential power himself in the year 2001, the second President Bush applied the lessons of Reagan's Vietnam Syndrome, peppered with a curious blend of Wilsonian messianism and realpolitik faith in military power, to invade Iraq and thereby restructure the politics of the Middle East, retrieve and extend American hegemonic credibility, establish military bases, and secure oil resources. In this, he was advised by radical neoconservative veterans of the 1960s who had served in the Nixon and Reagan administrations, and who were also militantly pro-Israel. Rejecting the anti-Vietnam War movement's critique of policymakers' hubris, George W. Bush's administration arrogantly blundered into an Iraqi quicksand war.

The Vietnam stab-in-the-back theory is so flawed in argument and so narrow in perspective that the real mystery is why it enjoyed and enjoys any acceptance. One of the reasons is that it embodies the ideals and institutions of powerful and influential segments of the society. The ambiguity of failure in both the War of 1812 and Korean War had made it possible for some influential US officials to overcome anti-civilian explanations of defeat with declarations of moral, political, or military victory, consequently preserving the American myth of omnipotence, or at least that part of the myth claiming America had never lost a war. The certain knowledge of defeat in the Vietnam War, however, like the certainty of the South's defeat in the Civil War, has helped bring into existence a myth of the Lost Cause, of which the stab-in-the-back legend is an essential part.

Key elements in the South's myth of the Lost Cause included stab-in-the-back themes, such as criticisms of Southern diplomats, civilian leaders, politicians, and others. General Lee, the central figure of the myth – the symbol of the superiority of Southern character, the invincible hero of the war – was portrayed as a tragic figure who, though superior to Ulysses S. Grant and other Northern generals, was defeated by the betrayals of the Confederate government and of subordinates like James Longstreet.[24] As Albion W. Tourgée observed about post-Reconstruction American fiction concerning Southern Civil War history, 'The downfall of empire is always the epoch of romance.'[25]

By the tenth anniversary of the end of the war in Southeast Asia, a Vietnam War version of the Lost Cause had arisen to become, as did the post-Civil War Lost Cause, a civil religion.[26] President Ronald Reagan ritualistically declared the war to have been a noble, honourable, idealistic, humanitarian Lost Cause. Veterans struggled with memory, frustration, loss, and the burden of defeat, while *Rambo*-like films symbolically refought the war and won (to popular acclaim), simultaneously exploiting veterans' anguish, POW mistreatment, and grief over those missing in action.[27]

Soldier-of-fortune magazines paid tribute to the memory of professional prowess and bravery, affirming the good cause, recommending the avoidance in future conflicts of mistakes made in Vietnam, and lamenting that soldiers had not been allowed to win[28] – at the same time publicizing mercenary camps and sending 'freedom fighter' instructors to Third World hot spots.[29] Right-wing evangelists bemoaned the abandonment of the war, the alleged fulfilment of the domino theory, and the victory of 'peaceniks'; they called for a redeeming revival of pre-1960s traditional values.

Former, allegedly weak presidents were ridiculed. Youngsters sported jungle-fatigue apparel.[30] Polls revealed that '18- to 22-year-olds were . . . among the most likely to share President Reagan's endorsement of the Vietnam War as a "noble cause"', and to believe it 'taught us that military leaders should be able to fight wars without civilian leaders tying their hands'.[31] War toys, war games, and the Reserve Officers' Training Corps (ROTC) made a comeback. Gun-ownership soared, justified, ironically, by a very loose interpretation of the Constitution (and encouraged in no small measure by political contributions from the National Rifle Association). Right-wing 'militias' sprung up in the name of Americans' right to defend family and home. Militarism was equated with patriotism. A new, nationalistic fervour called for vindication. Defeat was transformed into anger, which itself was transmuted into a desire to strike out – to win one for our side. National security officials, foreign policy committees, and conservative foundations and think tanks took encouragement from public-opinion polls said to show that Americans shifted from 'pacifism' to 'patriotism' in the post-Vietnam War era,[32] and they drew on popular sentiment about the war's Lost Cause to call for a return to militant containment policies, the rehabilitation of America's declining place in the world, and the re-establishment of American hegemony.[33] At stake for these former and present policy-makers was the issue of whether the Vietnam Syndrome would hinder containment of militarism and counter-revolutionary interventionism.

By the time of the presidential election of 2004, however, it seemed that Americans' historical memory of the Vietnam War had not only dimmed but had reached a state of equilibrium, in which the conflicts of the 1960s and the rancour of the war had been transformed into bittersweet nostalgic tolerance, and for the young, ancient history. But the nasty hoopla about the Vietnam War records of candidates George W. Bush and John F. Kerry during the 2004 campaign disabused us of this comforting notion. The forces of the Left and Right mobilized to refight the battles of the Vietnam War era, with the Right reprising the stab-in-the-back charge against Kerry, a genuine war hero (yet whose valour the Right slandered) who had returned home to oppose the war and thereby became an antiwar icon.[34]

The stab-in-the-back legend lives on as an inherent part of the foreign policy worldview of the Bush administration and its most ardent supporters. It is part of their historical memory of how the Vietnam War ended. It is also a core element in their Lost Cause of imposing the religious, evangelical Right's 'moral' and corporate agenda upon the rest of the nation, and of re-establishing, maintaining, and extending American hegemony. For kingmakers like Karl Rove, it is a means of defeating liberal challengers to the new Republican ascendency.[35]

Meanwhile, victorious Vietnam – so devastated by the war and its ironic diplomatic aftermath, when the United States and China colluded during the 1980s to punish Vietnam – is no longer the crucible of American anti-revolutionary, anti-communist policy. Although it is low on the horizon of Americans' vision, it is at the cusp of twenty-first century globalization, striving under a still-'communist' regime to advance on the road to modernization without falling victim to post-Cold War neocolonialism. The surviving leaders of the party and veterans of the army of North Vietnam and the Viet Cong of South Vietnam still struggle to understand the history of the war they won, but, while taking pride and satisfaction in their victory, they, or many of them, have been disillusioned by the disappointing outcome of so costly a victory. Except for their bitterness about the death, wounding, destruction, and environmental damage wrought by American military technology, many of the Vietnamese who remember the war, as well as the young, feel friendly towards the American people, many of whom they recall as having opposed the war, and, in any case, they are preoccupied with survival in the present and achieving prosperity in the future. The old veterans remain angry with the Chinese, who, they believe, stabbed them in their back, but pragmatists accept the necessity and appreciate the opportunities and dangers of having cordial diplomatic and commercial ties with China. In the shadow of the long-ago Vietnam War, it is the American Right – conceived broadly – that nurtures the stab-in-the-back legend and the ever-evolving Lost Cause.

Notes

1. Wolfgang Schivelbusch, *The Culture of Defeat: On National Trauma, Mourning, and Recovery*, trans. Jefferson Chase (New York: Henry Holt and Company, 2003), p. 15 (originally published in Germany in 2001 under the title *Die Kultur der Niederlage* by Alexander Fest Verlag, Berlin).
2. Jeffrey Kimball, 'The Stab-in-the-Back Legend and the Vietnam War', *Armed Forces and Society* 14 (Spring 1988), 433–58.
3. Schivelbusch, *Culture of Defeat*, pp. 15, 293–4, 304 n34.
4. In addition, several of his other criteria are of questionable relevance. Besides national polarization or civil war, national collapse or humiliation, the existence of a mythology of betrayal, counter-revolution, and true defeat, Schivelbusch also discussed the following stab-in-the-back criteria or symptoms: class and sociological clash between the stab-in-the-back accusers and their targets; the defeated nation's or group's claims that its defeat was the consequence of the victor's unmilitary means, its instinct for revenge, its martyr propaganda, its embrace of a glorious past, its faith in its cultural or moral superiority, and its defeat seen by as moral purification. Surely, Schivelbusch has created too many criteria, so many, it seems, that the only historical case of the stab-in-the-back phenomenon to meet his criteria would be the post-World War I German example. Nonetheless the post-Vietnam War example in the United States does analogously meet many or all of these criteria.
5. John E. Mueller, *War, Presidents and Public Opinion* (New York: John Wiley & Sons, 1973), Chap. 4, especially pp. 85–8, 90–1, 105–6.
6. Ibid., Chaps. 4 and 5.

7. Thomas C. Thayer, *War without Fronts: The American Experience in Vietnam* (Boulder: Westview Press, 1985), p. 105. For an analysis of public opinion polls on casualties, the economy, and prospects for a long war, see Mueller, *War, Presidents, and Public Opinion*, Chap. 3, especially pp. 36, 38, 56.
8. See, for example, Vietnam Policy Alternatives, [ca. 27 December 1968], folder10: Vietnam – RAND, box 3, HAK Administrative and Staff Files – Transition, Henry A. Kissinger Office Files, Nixon Presidential Materials (NPM), US National Archives.
9. Diary Entries, 19 April 1969 and 13 May 1971; Journals and Diaries of Harry Robbins Haldeman, NPM. Regarding Schivelbusch's criterion of civil war, there was no civil war on the scale of a true civil war such as the American Civil War of 1861–5 or the Russian Civil War of 1918–21 in at least two of the three comparative case studies he used to discuss the culture of defeat: the United States after the South's defeat at the hands of the North and Germany after World War I.
10. Mary Brennan, *Turning Right in the Sixties: The Conservative Capture of the GOP* (Chapel Hill: University of North Carolina Press, 1995), pp. 122–3. See also, Jules Witcover, *The Resurrection of Richard Nixon* (New York: G. P. Putnam's Sons, 1970), Chap. 6.
11. Thomas J. McCormick, *America's Half-Century: United States Foreign Policy in the Cold War and After* (2nd edn, Baltimore: Johns Hopkins University Press, 1995), pp. 155–65; Carole Fink, Philipp Gassert, Detlef Junker (eds), *1968: The World Transformed* (New York: Cambridge University Press, 1998); Jeffrey Kimball, *Nixon's Vietnam War* (Lawrence: University Press of Kansas, 1998), pp. 56–61.
12. Robert J. Graham, 'Vietnam: An Infantryman's View of Our Failure', *Military Affairs* 48, 3 (July 1984), 133–9.
13. Robert D. Heinl, Jr, 'The Collapse of the Armed Forces', *Armed Forces Journal* (June 1971), 30–7.
14. Arthur Marwick was probably the first to suggest the periodization of 'the long sixties'. Using mostly 'cultural' criteria, he dated the period from 1958 to 1974; *The Sixties Cultural Revolution in Britain, France, Italy, and the United States, c. 1958–c. 1974* (Oxford: Oxford University Press, 1998). I think the more relevant dates for the *American* experience of the 'long sixties' are 1956–75, from the Montgomery bus boycott, which began on 1 December 1955, to the fall of Saigon on 30 April 1975. On the 60s' roots of the 'counterrevolution', see, for example, Barry Werth, *31 Days: The Crisis That Gave Us the Government We Have Today* (New York: Doubleday, 2006).
15. Allen J. Matusow, *Nixon's Economy: Booms, Busts, Dollars, and Votes* (Lawrence: University Press of Kansas, 1998); McCormick, *America's Half-Century*, Chaps. 8 and 9; James Tybiel, *Ronald Reagan and the Triumph of American Conservatism* (New York: Pearson Longman, 2005), pp. 95, 96, 112–15, 120–1, 130.
16. He did not, however, clearly define what he meant by 'culture'. Moreover, he identified Freudian psychological processes as contributory causes rather than social, cognitive, and behavioral processes, which most psychologists now emphasize.
17. Memcon, National Security Council Meeting, 8 May 1972, box 998, Haig Memcons (January–December 1972), Alexander M. Haig Chronological Files,

NSC Files, NPM; Kimball, *Nixon's Vietnam War*, passim; and Kimball, *The Vietnam War Files: Uncovering the Secret History of Nixon-Era Strategy* (Lawrence: University Press of Kansas, 2004), Chaps. 5 and 7.

18. H. Bruce Franklin, *M.I.A. or Mythmaking in America* (New Brunswick: Rutgers University Press, 1993); Kimball, *Nixon's Vietnam War*, 165–75, passim; Kimball, *The Vietnam War Files*, pp. 29, 126, 161–2, 172; Memo, Holdridge to Kissinger, 20 March 1970, sub: Ambassador Habib's Presentations in Paris; and Memo, Kissinger to Nixon, 23 March 1970, box 3, WH/NSC: POW/MIA, NPM.
19. Jeremy M. Devine, *Vietnam at 24 Frames a Second* (Austin: University of Texas Press, 1995), p. 233.
20. George C. Herring, 'American Strategy in Vietnam: The Postwar Debate', *Military Affairs* 46 (April 1982), 57–62; Thomas G. Paterson, 'Historical Memory and Illusive Victories: Vietnam and Central America', *Diplomatic History* 12, 1 (Winter 1988), 1–18.
21. See, for example, Peter Steinfels, *The Neoconservatives: The Men Who Are Changing America's Politics* (New York: Simon & Schuster, 1979); Murray Friedman, *The Neoconservative Revolution: Jewish Intellectuals and the Shaping of Public Policy* (New York: Cambridge University Press, 2005); Francis Fukuyama, *America at the Crossroads: Democracy, Power, and the Neoconservative Legacy* (New Haven, CT: Yale University Press, 2006).
22. George C. Herring, 'The "Vietnam Syndrome" and American Foreign Policy', *Virginia Quarterly Review* 57, 4 (1981), 594–612; Walter LaFeber, 'The Last War, the Next War, and the New Revisionists', *Democracy* 1, 1 (January 1981), 93; Michael T. Klare, *Beyond the 'Vietnam Syndrome': US Interventionism in the 1980s* (Washington, DC: Institute for Policy Studies, 1981); Kimball, 'Stab-in-the-Back Legend'; Robert Buzzanco, 'Fear and (Self-) Loathing in Lubbock, Texas, or How I Learned to Quit Worrying and Love Vietnam and Iraq', *Passport: Newsletter of SHAFR* 36, 3 (December 2005), 5–12.
23. *Public Papers of the Presidents: Ronald Reagan, 1981*, 155–6; *1982*, p. 1445; *1983*, p. 255; Laurence I. Barrett, *Gambling with History: Ronald Reagan in the White House* (New York: Penguin Books, 1984), p. 41; 'President Promises', *Journal-News* (Hamilton-Fairfield, Ohio, 20 July 1986), A-1.
24. Thomas L. Connelly, *The Marble Man: Robert E. Lee and His Image in American Society* (New York: Alfred A. Knopf, 1977).
25. Albion W. Tourgeé, 'The South as a Field for Fiction', *Forum* 6 (1888–9), 412.
26. In *Baptized in Blood: The Religion of the Lost Cause, 1865–1920* (Athens: University of Georgia Press, 1980), Charles Reagan Wilson argues that the Lost Cause became a civil religion – a set of semi-institutionalized public beliefs and values embodying historical mythology.
27. *Rambo: First Blood Part II* (1985); *Missing in Action* (1984). Some 1980s motion pictures tapped Vietnam War stab-in-the-back and Lost Cause themes even though their immediate subject matter was not about Vietnam; for example, *Red Dawn* (1985) and *Amerika* (1987).
28. See *Vietnam Combat* (Spring 1986) and *Back to Battle: Vietnam* (pulp magazines published by *Soldier of Fortune*, February 1986).
29. 'Militarism in America', Centre for Defense Information, *The Defense Monitor* 15, 3 (1986), 7.
30. 'Costume Party!' *J.C. Penney's Christmas 1986 Catalogue*, pp. 403, 410–11, blends sexism, racism, and militarism.

31. Adam Clymer, 'What Americans Think Now', *New York Times Magazine* (31 March 1985), 34.
32. Jerry W. Sanders *Peddlers of Crisis: The Committee on the Present Danger and the Politics of Containment* (Boston: South End Press, 1983), p. 193.
33. Christopher Hitchens, 'Home Is Where the Enemy Is: Oliver North's "Lessons" of Vietnam', *In These Times* (25–31 March 1987), 12–13; and Oliver L. North testimony before the Select Committee on Secret Military Assistance to Iran and the Nicaraguan Opposition, 9 July 1987, in *Taking the Stand: The Testimony of Lieutenant Colonel Oliver L. North* (New York: Pocket Books, 1987), pp. 269–71. Portions of the last three paragraphs were adapted from my article, 'Stab-in-the-Back Legend'.
34. See, for example, 'Kerry Pressing Swift Boat Case', *New York Times* (28 May 2006).
35. In an article published six months after the Edinburgh conference, novelist Kevin Baker traced the American stab-in-the-back legend from right-wing criticisms of FDR, through the Vietnam War, to the present: 'Stabbed in the Back! The Past and Future of a Right-wing Myth', *Harper's Magazine* (June 2006).

Index